Make It, Wear It

Wearable Electronics for Makers, Crafters, and Cosplayers

Make It, Wear It
Wearable Electronics for Makers, Crafters, and Cosplayers

Sahrye Cohen and Hal Rodriguez

New York   Chicago   San Francisco   Athens   London
Madrid   Mexico City   Milan   New Delhi
Singapore   Sydney   Toronto

Library of Congress Control Number: 2018942700

Make It, Wear It: Wearable Electronics for Makers, Crafters, and Cosplayers

1 2 3 4 5 6 7 8 9   LWI   23 22 21 20 19 18

ISBN   978-1-260-11615-1
MHID   1-260-11615-8

Sponsoring Editor	**Acquisitions Coordinator**	**Indexer**
Robert Argentieri	Elizabeth Houde	Jack Lewis
Editing Supervisor	**Project Manager**	**Art Director, Cover**
Donna M. Martone	Patricia Wallenburg	Jeff Weeks
Production Supervisor	**Proofreader**	**Composition**
Lynn M. Messina	Alison Shurtz	TypeWriting

About the Authors

Sahrye Cohen is the co-founder and chief designer of the tech couture design group, Amped Atelier. She teaches workshops on wearable electronics for costumers and cosplayers and has published articles on 3D printing in *Make:* magazine and on cosplay techniques in *The Virtual Costumer*.

Hal Rodriguez is a maker and programmer based in the San Francisco Bay Area. He is the co-founder and chief technologist of the tech couture design group, Amped Atelier. He has over 20 years of programming experience and has published articles on 3D printing in *Make:* magazine.

Contents

Acknowledgments xi

1 Introduction **1**
The Projects 4

2 Supplies, Techniques, and Fabrication Machines **7**
Sewing Supplies 7
Hand and Machine Stitches 10
Patterns 12
Electronic Tools and Components 13
Electronic Techniques 17
Sewing with Conductive Thread 20
Microcontroller Boards 22
Laser Cutters 22
Laser Cutting Supplies 24
Design for the Laser 24
3D Printing 25
3D Design and Printing 26
Using Tinkercad 26
Slicing with Cura 28
Downloadable Files for This Book's Projects 29

3 3D Embellished T-Shirt **31**
What You Will Learn 31
Files You Will Need 31
Tools You Will Need 31
Materials You Will Need 32

Step 1: Cut Out the Shirt Pieces   34
Step 2: Create Your 3D Design   34
Step 3: Export and Slice the Model.   35
Step 4: 3D Print on Fabric.   37
Step 5: Assemble the T-Shirt   40

4   Fiber-Optic Fabric Scarf **45**

What You Will Learn.   45
Files You Will Need.   45
Tools You Will Need   45
Materials You Will Need   46
Step 1: Laser Cut the Fabric for the Scarf.   48
Step 2: Split the Fiber-Optic Panel   48
Step 3: Sew the Scarf Fabric and Attach It
        to the Fiber-Optic Panel   52
Step 4: 3D Print the Battery and LED Holder   55
Step 5: Attach the Light Source   55
Going Further   57

5   Fun Festival Hip Pack. **59**

What You Will Learn.   59
Files You Will Need.   60
Tools You Will Need   60
Materials You Will Need   60
Step 1: Program the Circuit Playground Express   61
Step 2: 3D Print the Circuit Playground Express Cover   63
Step 3: Make the Pack   63
Step 4: Bring It All Together   70

6   Solar Backpack **73**

What You Will Learn.   73
Files You Will Need   73
Tools You Will Need   73
Materials You Will Need   74
Step 1: Prepare the Solar Charger   75
Step 2: Prepare the Solar Panel   77
Step 3: Solder the USB Connector on the Power Boost 500.   78
Step 4: Splice Two JST-PH Connectors End to End.   78
Step 5: Hook Everything Up.   80
Step 6: Make the Backpack.   82
Step 7: Attach the Solar Panel to the Backpack   97
Step 8: Use It Outside   99

7 Starlight Skirt **101**

What You Will Learn.. 101

Files You Will Need... 101

Tools You Will Need 101

Materials You Will Need 102

Step 1: Sew the Skirt 103

Step 2: 3D Print the Fiber-Optic Holders and LED Diffusers..... 113

Step 3: Put the Fiber Optics Together 114

Step 4: Connect the Electronics............................. 119

Step 5: Program the StitchKit's MakeFashion Board........... 121

Step 6: Going Further 124

8 Programmable Sewn Circuit Cuff.......................... 127

What You Will Learn.. 127

Files You Will Need... 127

Tools You Will Need 127

Materials You Will Need 128

Working with Leather 130

Step 1: Make the Cuff 131

Step 2: Charge the Battery.................................. 131

Step 3: Lay Out and Sew the Electronics...................... 132

Step 4: Program the Gemma 134

Step 5: Use the Capacitive Touch Sensor 136

Step 6: Print a Battery Case................................. 138

Step 7: Going Further 138

Step 8: Troubleshooting.................................... 139

9 LED Matrix Purse....................................... 141

What You Will Learn.. 141

Files You Will Need... 141

Tools You Will Need 141

Materials You Will Need 142

Step 1: Laser Cut the Purse Case and Mounting Panel 145

Step 2: Update the Firmware................................ 145

Step 3: Prepare the PowerBoost 1000 146

Step 4: Attach the Electronics to the Mounting Panel 147

Step 5: Assemble the Case.................................. 150

Step 6: Use It!... 155

Index... 157

Acknowledgments

THANK YOU TO THE EDITORS AND STAFF AT McGRAW-HILL TAB FOR YOUR INTEREST AND professionalism in guiding this book to completion. Fantastic maker Alison Lewis provided some excellent advice at the start of this project; and Shannon Hoover, Maria Elena Hoover, and the community at MakeFashion really helped us get started making fashion tech. Our wonderful models, Jade Rose, Monica Jackson, Ky Faubion, and G.G., brought the projects beautifully to life. Many thanks to Jason Martineau for his artistry and wonderful photography. Thanks also to Galeet Cohen, Talia Timmins, and Holly Costa for help in organizing, proofreading, and general encouragement.

Make It, Wear It

Wearable Electronics for
Makers, Crafters, and Cosplayers

CHAPTER 1

Introduction

WITH WEARABLE TECHNOLOGY, YOU CAN MAKE FUNCTIONAL, USEFUL, AND BEAUTIFUL GARMENTS that become an interface, connecting people to other people and promoting meaningful interactions between individuals and the whole of the world around them. Clothes become another part of the digital world, interacting with smartphone applications and social media views and giving onlookers a glimpse into your personality, your mood, and even your thoughts. In this book we've designed projects that combine electronics with modern fabrication technology such as 3D printers and laser cutters to produce wearable projects that a few years ago were only accessible to major designers and fashion houses. Using home machines and shared makerspaces, you can make personalized clothing and accessories that bring the imaginings of a science fiction future to today.

Electronics, whether a simple light-up circuit or a complex programmed design, can create visual interest and fun interactive clothing. Want to play video games directly on your clothes? The Gamer Girls dresses by Phi Designs uses programmable lights for two players to play against each other right on the front of the dresses (Figure 1.1).

Other clothes can interact with social media, using smartphone applications and microprocessors to display words from Twitter or other social platforms. Reflections, designed by Amped Atelier, uses a smartphone app to scroll words on the front of an evening gown (Figure 1.2).

Figure 1.1   Gamer Girl dresses by Phi Designs use microcontrollers and addressable light-emitting diodes (LEDs) to play a video game right on the wearer's dress. (*Courtesy of Ernesto Augustus at the Make Fashion Gala, 2016*)

Figure 1.2   Connecting LEDs with a smartphone app allows clothes to display words, animations, and even social media connections. (*Courtesy of Rafal Wegiel at the Make Fashion Gala, 2015*)

Using the tutorials in this book, you will be able to make your own projects combining electronics and basic sewing to make clothes that glow and twinkle. Precision cutting and engraving using laser cutters creates intricate, detailed decoration and rapid fabrication of fashion-forward accessories.

The projects in this book will guide you through using laser-cut fabric as a design element, as well as using the laser to make a modern acrylic clutch purse. 3D printing makes tabletop production of components and tools easy and affordable. 3D printers can also be used to make fantastic decorations and light-up embellishments (Figure 1.3).

Figure 1.3   3D printing creates futuristic light diffusers and unique embellishments. (*Courtesy of Jeff McDonald at the Make Fashion Gala, 2016*)

In addition to using 3D printers to make electronics enclosures, these projects will use 3D printing to make accessories and produce personalized clothes (Figure 1.4).

Figure 1.4 Wearable technology makes functional and beautiful garments that can interact with the wearer and the environment. (*Courtesy of Maria Elena Hoover at the Motion Ball, 2018*)

The Projects

Maybe you've seen 3D printers and wondered how they can be useful for wearable projects. Or maybe you make your own clothes and want to add modern glitz and interaction. If you are a maker, programmer, costumer, cosplayer, hacker, engineer, or someone who wants to experiment and learn, this book is for you. The projects are designed to combine the advantages of rapid fabrication using 3D printers and laser cutters with basic sewing construction. Creating wearable electronics fashion requires using skills from sewing, programming, computer-aided design (CAD), and many more disciplines. This book will provide a basic introduction to wearable technology with the goal of completing the specific projects and inspiring creativity. There are many fantastic resources available from this publisher and others to help develop your skills in sewing and garment construction and to learn more about the intricacies of electronics and programming. For projects that require sewing an entire garment or accessory, we've provided no-sew or low-sew alternatives for incorporating the electronics and fabrication techniques onto purchased clothing and bags.

This book will take you step-by-step through making projects using the 3D printer and laser cutter and building electronics skills. Each project will use information from the introductory chapters and the preceding projects to guide you from constructing simple electronic circuits to using microcontrollers such as the Arduino while introducing major programming methods including visual block programming and the Arduino IDE.

Wherever possible, we've provided electronic downloads that are available from the book website with clothing and accessory patterns, CAD files for laser cutting, 3D printing files, and microcontroller programs to help you complete your projects. The projects use a variety of different materials, including knit fabric, polylactic acid (PLA) plastic, acrylic, and even fiber-optic fabric. When you understand how to make wearable projects, you can experiment with different materials and mix-and-match techniques, combining parts of projects together to make the perfect personalized clothes and accessories.

CHAPTER 2

Supplies, Techniques, and Fabrication Machines

THIS CHAPTER PROVIDES A BRIEF INTRODUCTION TO SEWING SUPPLIES AND TECHNIQUES AND an overview of basic electronic components and techniques. It also provides necessary information on computer-aided design/computer-aided manufacturing (CAD/CAM) and using laser cutters, 3D printers, and microcontrollers.

Once you learn to sew, the clothing you can create is limitless. We will be using hand sewing to sew electronic components onto garments and also machine sewing to construct garments, bags, and accessories. If you are new to hand or machine sewing, you may want to watch some introductory *How to Sew* videos on YouTube and consult your sewing machine manual before you get started. The fabric that you use to make clothes or accessories should be washed and ironed before you cut it out and use it for your project.

Sewing Supplies

Fabric Scissors

Fabric scissors are usually metal and have blades that are between 5 and 8 inches long. Try not to cut paper with your fabric scissors because paper will dull the blades.

Pins

Straight sewing pins will help keep your fabric pieces together as you sew by hand or with a sewing machine. Safety pins can also be used to help keep your projects together while you are working on them.

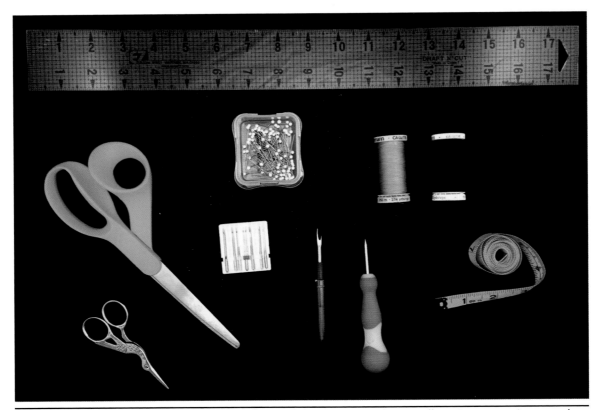

Figure 2.1   Basic sewing supplies (from left to right): scissors, sewing needles, pins, seam ripper, awl, thread, measuring tape, and ruler (top).

Sewing Needles

Hand sewing needles come in a variety of sizes. Use finer needles for lighter fabrics and thicker needles for heavy fabrics. A size 7 needle should be appropriate for most projects in this book. A large-eye needle should be used for sewing conductive thread because the thread is thick.

Thread

All-purpose polyester thread should be used to construct the garments and accessories.

Seam Ripper

This small tool has a sharp edge that rips out stitches.

Measuring Tape

A flexible measuring tape will help you measure your waist and chest to decide which size to make. If you don't have a measuring tape, you can use a piece of string and measure it with a ruler.

Ruler

A clear, bendable 18-inch ruler will make measuring and marking very easy.

Awl

A sewing awl has a sharp point that makes a hole in fabric by spreading fibers apart instead of cutting. This will help to ensure that your fabric doesn't unravel.

Leather Needle

This is a special sharp needle with a triangular point for sewing through leather. Leather needles can be purchased at leathercraft stores, many major craft stores, and some hobby stores.

Iron

A basic iron that can press natural and synthetic fabrics will help flatten seams and keep your project neat as you sew.

Sewing Machine

A sewing machine that can do a straight stitch, a zigzag stitch, and a backstitch will allow you to sew every project in this book. An 80/12 ballpoint needle should be used for T-shirt knits, a heavier 100/16 or 110/18 needle is used for fabrics such as canvas or denim, and a medium-sized 90/14 needle is appropriate for most woven garment fabrics. A serger is a specialized sewing machine that makes an overlock stitch and a coverstitch, like the seams and hem on your T-shirt. A serger is not necessary for the projects in this book but may be used if you have one.

Hand and Machine Stitches

Running stitches and whipstitches are the primary stitches we will use for our projects. All hand stitching starts by threading a needle and knotting the far end of the thread so that it can't be pulled all the way through the fabric. A running stitch is used to sew two pieces of fabric together. You create it by inserting a needle into and out of the fabric several times in a line of small stitches (about $\frac{1}{8}$ to $\frac{1}{4}$ inch long). A whipstitch is used on buttonholes to help keep the fabric from unraveling and can also be used to sew electronic components and covers onto fabric. Push the needle through the fabric near the edge. Wrap the thread around the edge of the fabric, and push the needle through again next to your first stitch. This technique will also be used to sew components onto the fabric. Take a small stitch completely through the front and back of the fabric. Wrap the thread around the component or cover, and then take another stitch completely through the fabric. Use more stitches until the component is securely attached. Then knot the thread to finish (Figure 2.2).

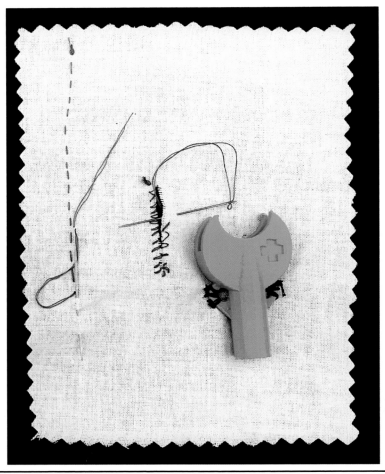

Figure 2.2   Hand-sewing techniques (from left to right): running stitch, whipstitch on edge, and whipstitch attaching components to fabric.

The machine stitches we will use are the straight stitch, the zigzag, and the backstitch. Straight stitches are the basic stitch used to join two pieces of fabric together. A zigzag stitch moves the needle left and right as you sew. It has more elasticity and is used for sewing seams on knits. It can also be used to help keep fabric edges from unraveling. A backtack, or small backstitch, is the small stitch forward and back that you should use at the beginning and end of every seam in order to keep your sewing from coming undone (Figure 2.3).

Your sewing machine manual is a good reference to learn how to do each of these stitches. A seam allowance is the correct distance from the edge of your fabric to place your seam. Most projects in this book will use a standard ½-inch seam allowance.

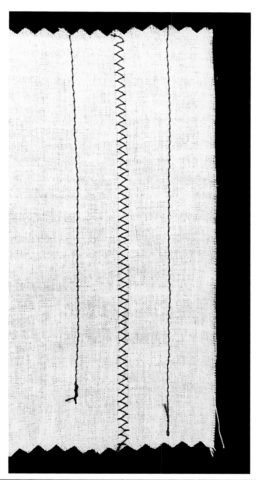

Figure 2.3 Machine sewing techniques (*from left to right*): straight stitch, zigzag stitch, and straight stitch with a ½-inch seam allowance.

Patterns

The downloadable associated materials for this project include a raglan-sleeve T-shirt pattern and a backpack pattern that you will be able to print on paper using a home printer. If you are using 8½- × 11-inch paper, you may need to print several sheets and tape them together to get the full pattern piece.

The first step in making accurate garments is to carefully cut out the paper patterns. After cutting out the paper, spread your fabric on a hard, flat surface. Place all the paper pattern pieces on your fabric, paying attention to instructions such as "cut on fold" or "cut two." Pin the paper pieces to your fabric to ensure that they will stay in place as you use scissors to cut the fabric pieces out along the lines of the paper pattern (Figure 2.4).

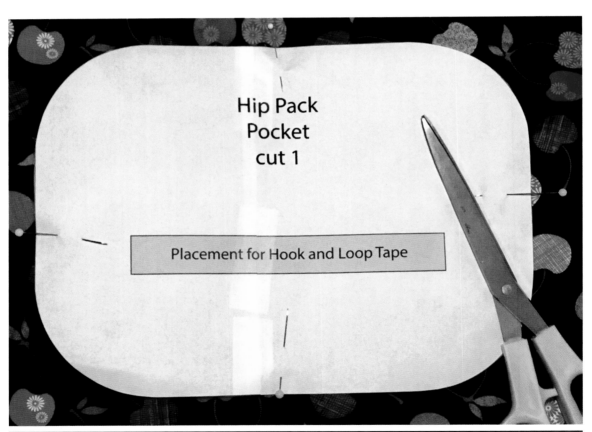

Figure 2.4   Paper pattern pinned to fabric before the fabric is cut out.

Electronic Tools and Components

Electronic tools and components are an important part of the fashion tech maker's toolbox (Figure 2.5). While some of these tools and supplies may be available at local hobby shops and electronics supply stores, most will need to be ordered online. We have sourced the materials for these projects from well-known and reliable online stores, including Adafruit (www. adafruit.com), SparkFun (www.sparkfun.com), and Seeed Studio (www.seeedstudio.com).

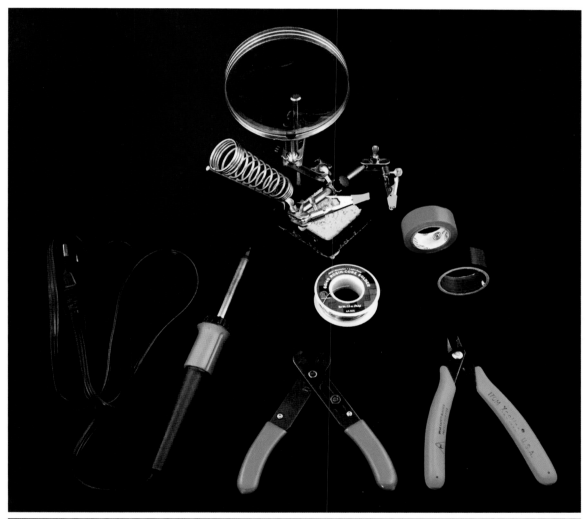

Figure 2.5   Electronic tools and supplies.

Light-Emitting Diodes

A light-emitting diode (LED) lights up when current flows through it. LEDs are polarized, meaning that the current from your battery can only run through them in one direction, from the positive side (called an *anode*) to the negative side (called a *cathode*). In a simple LED with two legs, the longer side is the positive side; other LEDs may have the positive side marked with a plus (+) sign.

Addressable LEDs

LEDs can be individually set to a certain color by varying the levels of red, green, and blue light to make many different colors. These LEDs have a chip that receives commands from a controller, such as an Arduino. Addressable LEDs are available individually, on strips or rings, and as a matrix (Figure 2.6). Common versions are the WS2812B and also any with the brand name NeoPixel.

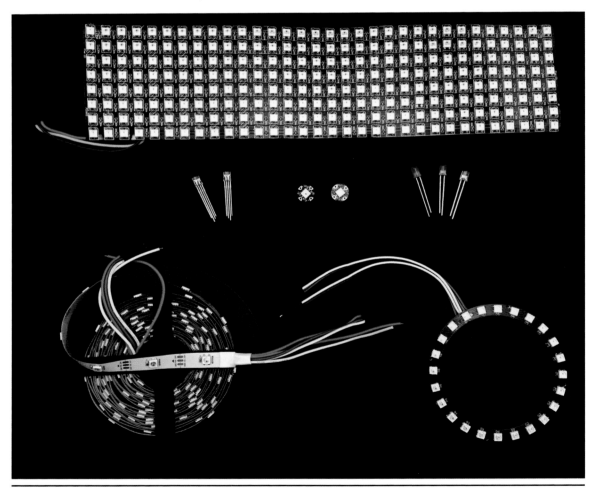

Figure 2.6 Simple LEDs and addressable LEDs are made individually, as strips or rings, and as a matrix.

Battery

Batteries store the energy used for electronic circuits. In this book we will use small coin cell batteries (CR2032), lithium polymer (LiPo) batteries, and large batteries that are commonly available as cell phone chargers. LiPo batteries are used a lot in wearable projects but can be dangerous if mishandled. Never pierce, bend, or overheat a LiPo battery because it can catch on fire.

Wire

Electrical wire has a conductive metal material on the inside to carry electric current and insulation and a nonconductive material on the outside. Solid-core wire is one solid strand of a particular gauge measured by the American Wire Gauge (AWG) standard. Stranded wire contains multiple strands inside the insulation and bends more easily. Silicone-covered wire is recommended for wearable projects because it is very flexible.

Solder

Solder is a metal alloy that is melted to create a bond between two pieces of metal. For these projects, we will be creating a bond between a wire and the electronic component with solder, and we will also be using through-hole soldering, which attaches components to a circuit board. Typically solder is 60 percent tin and 40 percent lead (60/40), although lead-free versions are also available.

Soldering Iron

A soldering iron is the tool that melts solder to join your components. For these projects, a soldering iron that is at least 25 watts (W) and has a small tip is recommended for detailed work and small connections.

Conductive Thread

This special thread is sewn by hand or with a sewing machine, but it can conduct electricity like wire. Conductive thread is usually stainless steel or silver metal that is bonded to nylon or another fiber. It can be purchased online from electronics suppliers.

Electrical Tape

Insulating tape is used to wrap wires to protect components. When taping wires, be sure that exposed wires aren't touching other conductive surfaces.

Wire Cutter

This is a special tool used for cutting wires. Please note that cutting wire with scissors will dull the scissor blades.

Wire Stripper

This tool removes the plastic or silicone coating from electrical wires so that you can solder the bare wire underneath.

Safety Glasses

Always wear goggles or glasses when soldering because hot solder may splatter.

Helping Hands

Helping hands is a stand with multiple flexible clips that hold your components steady for soldering. This tool will be useful for some of the projects in the book that require soldering small boards and components.

Fiber Optics

These are thin, flexible fibers, usually made of glass or plastic, that transmit light. Fiber optics come as end-glow or side-glow fibers.

Microcontroller Boards/Arduino

A microcontroller board is an integrated circuit and supporting hardware that executes instructions typically for a specific process or set of devices. The Arduino is a common, inexpensive, easy-to-use open-source board. Other beginner and budget-friendly microcontroller boards for wearables are made by Adafruit, Sparkfun, Seeed Studio, and MakeFashion (Figure 2.7).

Sensors

Sensors are electronic components that measure some aspect of the physical world. Sensors can sense such things as light, temperature, touch, and movement. In combination with microcontrollers, sensors can measure the physical environment and display the information in various ways, including lights, colors, and sounds.

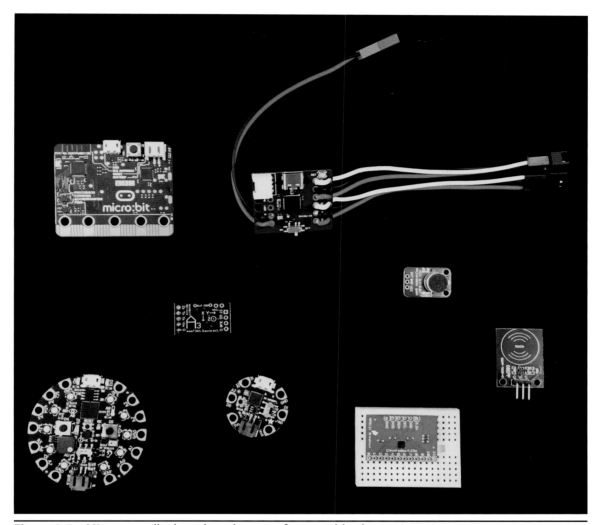

Figure 2.7 Microcontroller boards and sensors for wearable electronic projects.

Computer

A computer with an Internet browser will be needed to download project patterns and files and to access the programming interfaces and libraries for various microcontrollers. You may also need a computer to access CAD software for both laser cutting and 3D printing.

Electronic Techniques

Learning to make circuits and assemble electronics is an important skill set to acquire if you want to make your own wearable projects. Making a clean solder connection to connect wires together and soldering electronic components to a board are the most fundamental skills. There are many resources available from TAB Electronics at McGraw-Hill, including the *Evil Genius* series and *Teach Yourself Electricity and Electronics*, which will provide step-by-step projects to learn many electronics techniques. Watching online video tutorials and practicing with scrap wire and learn-to-solder projects are also great ways to build these essential skills.

To make the projects in this book, you will need to know how to solder wires together and also how to do through-hole soldering. Using a clean soldering tip, setting your soldering iron to a medium temperature (325–375 degrees Celsius), and wearing safety glasses are best practices. If you see smoke coming from your solder, try turning the temperature down or cleaning your tip using a brass sponge.

Soldering Wires Together

To prepare wire for soldering, you need to use your wire strippers to remove between $\frac{1}{4}$ and $\frac{1}{2}$ inch of insulation on the end, leaving the wire inside intact. Using a hot soldering iron, tin the ends of the wire by adding a little hot solder to each wire individually. This will help them to bond more easily when you are ready to join them together. Next, hold the two pieces of wire close together using pliers or a helping hands clamp. Touch the tip of the soldering iron to the point on the wire that you would like to connect. Hold the solder so that it is touching both the tip of the soldering iron and the wire at the connection point. Once the solder starts melting, you can use the tip of the soldering iron to spread the melting solder over the connection point between your two wires (Figure 2.8).

Through-Hole Soldering

The key to successful through-hole soldering is to heat the component, the board hole, and the solder at the same time. Insert the leg of your electronic component into the correct hole on the board. Touch the hot soldering iron tip to the board hole and the electronic component, and then add the solder until you see the solder start to melt. Once some solder has melted and stuck to the components, remove the rest of the solder and use the hot tip to melt the remaining solder through the hole and around the component leg. A good solder joint flows around the hole and the leg of the component, forming a little cone shape (Figure 2.9).

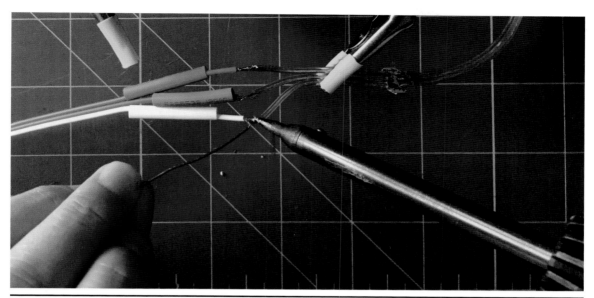

Figure 2.8 Soldering two wires together.

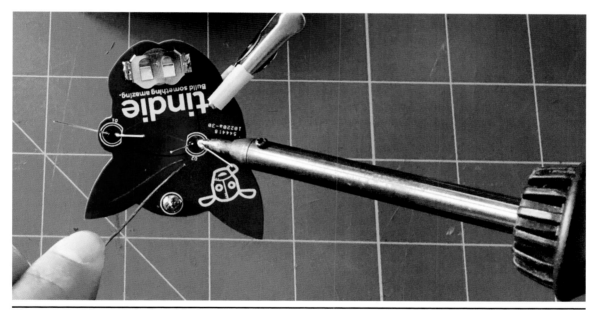

Figure 2.9 Through-hole soldering.

Sewing with Conductive Thread

Conductive thread can be sewn by hand using a regular sewing needle with a large eye. Because the thread can sometimes be a bit thicker than regular thread, you should cut the end you are threading at an angle, and you might want to use a needle-threader tool to help pull the thread all the way through. If you are sewing by hand, use a running stitch with the conductive thread to join your electronic components together through the fabric. Conductive thread is stiffer than regular thread and has more of a tendency to unravel. To keep your connections together, we recommend that you knot your thread, pass your knotted thread through the fabric, sew back through the thread just under the knot to secure it, and then start your stitching with several backstitches. To do a backstitch, you pass your knotted thread through the fabric and take two forward running stitches. You then take one stitch backward in between your two running stitches. Then take two more running stitches and backstitch again before continuing on with a regular running stitch (Figure 2.10). Adding a small dot of fabric glue, Fray Check, or clear nail polish to the knot can also help to keep your knots in conductive thread tight.

When sewing conductive thread to electronic components, it is important to get a good connection to the copper. First, pass your knotted thread through your fabric, and make one full backstitch to secure it. Then whipstitch tightly around the copper pad several times, even tying a small knot before continuing with your running stitch. To reduce the strain on your conductive thread, we recommend that you attach sewn components using a whipstitch with regular thread in addition to the conductive thread (Figure 2.11). Because conductive thread is not insulated, it is easy to accidentally create shorts by crossing the lines of the conductive thread. Carefully plan your sewing so that you don't have to cross lines of stitches, and when cutting knots, be certain that you aren't leaving long tails that could touch other conductive parts of the project.

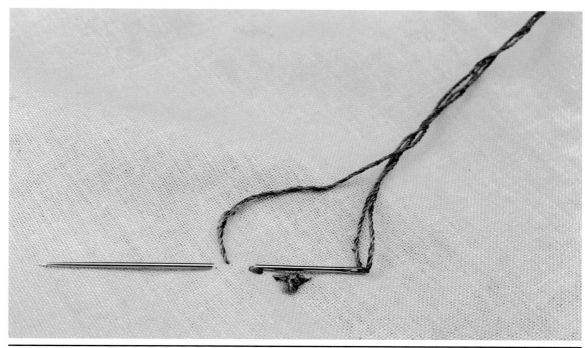

Figure 2.10   Backstitch with conductive thread by taking two running stitches and one backstitch.

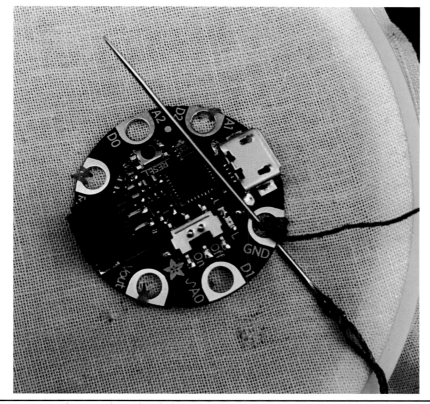

Figure 2.11   Wrap conductive thread tightly around a component with a whipstitch.

Microcontroller Boards

Microcontroller boards are the heart of interactive wearable projects. These boards typically perform one specific job using sensors, lights, or other inputs and outputs. Microcontrollers can do complex jobs, such as running a robot, but they depend entirely on the instructions you program. One of the most well-known boards is the Arduino (www.arduino.cc). This board is intended to be a learning tool, and the circuitry and design are open source, meaning that anyone can copy, modify, and sell circuit designs based on the original design. An all-in-one programming tool, the Integrated Development Environment (IDE), is typically used to program the Arduino and boards that are based on the Arduino. Because the design is so popular, there are fantastic resources available for learning how to program using the Arduino IDE.

Other easy-to-learn microcontroller boards, such as the micro:bit and Adafruit's Circuit Playground Express, use visual block programming to tell your board what to do. We will be using Microsoft MakeCode (makecode.com) for visual block programming. MakeCode has a growing number of microcontroller boards that can be used and a well-organized website with tutorials and learning resources.

Another increasingly common programming language for wearable projects is MicroPython, based on the popular programming language Python. The advantage of MicroPython is that it is easy to get started on low-cost microcontroller boards, and it does not require some extra download steps that are needed on some very small Arduino boards. CircuitPython is an extension of MicroPython with support for Adafruit products. We've provided programming shortcuts for the projects as downloads to get you started, but once you learn the methods, your creativity is unlimited.

Laser Cutters

A laser cutting machine works by directing a beam of focused light that is capable of cutting through a variety of materials (Figure 2.12). This computer-controlled beam allows for precision cutting and fine etching on thin materials, usually up to $\frac{1}{4}$ inch thick. The most common laser cutters use a carbon dioxide (CO_2) laser that burns and vaporizes to etch and cut through material. Natural and polyester fabrics, leather, thin wood, felt, acrylic, and many other materials can be cut on CO_2. Most readily available lasers cannot cut metals, glass, or potentially toxic materials such as vinyl and polyvinyl chloride (PVC).

Laser cutters can commonly be found at makerspaces and increasingly at colleges, schools, and libraries. There are even desktop laser cutters for home use made by a number of companies, including Epilog, Orion, and Glowforge. If you don't have a laser cutter available to you, there are local and online services, such as Ponoko (www.ponoko.com), that you can hire to cut your material from a computer file that you provide. An online search should locate makerspaces and custom cutting services in your area. You will usually be required to take a safety class before operating a laser cutter at a shared space. Always carefully follow the instructions and materials restrictions for the laser you are operating. Operating a laser incorrectly can quickly start fires or release toxic gases.

Figure 2.12   Laser cutters can cut and finely etch thin materials.

Laser Cutting Supplies

Natural and Manufactured Fabrics

Natural fabrics such as cotton, wool, and linen can all be cut with most lasers. Polyester, nylon, and other synthetic fabric blends cut well with slightly melted, sealed edges; some makerspaces may have restrictions on laser cutting synthetic fabrics. Both wool felt, and wool and synthetic blend fiber felt can be cut with a laser cutter. Do not cut fabrics with PVC because doing so can release toxic gases and also harm laser components.

Leather

Vegetable-tanned leather is an excellent material for laser etching and engraving. Some faux leathers or "vegan" leather contain PVC and should not be cut with a laser cutter.

Acrylic

Acrylic is a common plastic that is also known as Lucite, plexiglass, or Perspex. Acrylic comes in cast or extruded sheets, and ¼-inch thickness is a common size for laser-cut projects.

Design for the Laser

CAD software can create two-dimensional (2D) drawings or three-dimension (3D) models. When cutting material on a laser cutter, the machine is using a 2D vector graphic format. There are many pieces of software that can produce such files, including Inkscape (free and open source), Adobe Illustrator, CorelDRAW, and AutoCAD. Vector files usually have a file extension of .ai, .svg, .eps, .dxf, or .dwg. Which extension you will use depends on your laser cutter software; most CAD software will be able to convert or export vector files to different types. To get you started, the vector files for projects in this book are available as a download. Once you are comfortable using a specific CAD software, you can use the downloadable project files as a starting place to experiment and create personalized designs for electronics projects.

3D Printing

3D printing is a manufacturing process that makes a three-dimensional object from a digital design file by adding material together. By using a 3D printer, you can rapidly make individual, personalized parts for projects and also try out new designs without having to order many pieces from a commercial supplier. The most common home machines use very thin layers of plastic to slowly build a solid object. This is commonly called *fused-filament fabrication* because the printers use a roll of plastic filament as the construction material (Figure 2.13). Other 3D printing processes are resin printers, which can be found at professional printing services or in specialty home machines. Resin printers use photopolymerization (which uses light to cure a liquid resin) to create a 3D object or granular materials binding (which use lasers or other energy sources to fuse layers of powder together).

Figure 2.13   Filament 3D printer, filament roll, and printed objects.

3D printers are available at many makerspaces, public libraries, and schools; there are also online services, such as Shapeways (www.shapeways.com) that will 3D print your object in a variety of different materials. Most shared makerspaces will require a safety and basic-use class before you are cleared to use the 3D printer. We recommend that you use polylactic acid (PLA) filament for the projects in this book because it is widely available in many fantastic colors (even metallic and glow-in-the-dark filaments) and can be used in a wide variety of printers. PLA filament is also affordable and easy to use for beginning makers, and it is a bioplastic, meaning that it is made from renewable plant resources. If you can't purchase 3D filament from your makerspace, you can buy it online via some 3D printer manufacturers' websites and 3D printing supply sites such as MatterHackers (www.matterhackers.com). When choosing a personal 3D printer, you will want to consider the types of plastic filaments you want to use, the size of your print, print quality, and, of course, price. *Make:* magazine (makezine.com) publishes an annual review of 3D printers that can guide you in purchasing the perfect 3D printer for your needs.

3D Design and Printing

There are many different 3D modeling tools you can use to create original designs or to modify existing 3D models. Like 2D designs, these tools use CAD to produce a digital file. Some common software programs for 3D design include Tinkercad and Morphi App, which are great for beginners, and Blender and Fusion 360, which have more functionality once you learn the basics. All these programs have online tutorials and videos that will help you to get started making 3D designs.

The easiest way to get started with 3D design for printing is to look at the many existing designs already available online. You can download 3D designs from Thingiverse (www.thingiverse.com) and YouMagine (www.youmagine.com), which have thousands of free 3D designs. Experimenting with existing files will help you understand the capabilities of your printer and can also provide a base model to remix for building your own projects.

Using Tinkercad

Tinkercad (www.tinkercad.com) is a free browser-based 3D design and modeling tool. Tinkercad is very intuitive and easy to learn, especially if you don't have much experience with these tools. If you already have a favorite 3D design and modeling tool, you can use that.

Working in Tinkercad is easy. You can easily create simple designs using shape and text generators. If you don't already have a Tinkercad or Autodesk account, you will first need to create one.

Before you start designing, change the dimensions of the work plane to match either the size of the printer bed or, if the piece is larger than the printer bed, the size printing. You can change the work-plane dimensions by clicking "Edit Grid" in the lower left of the main Tinkercad view (Figure 2.14). You can also change the units in which dimensions are displayed.

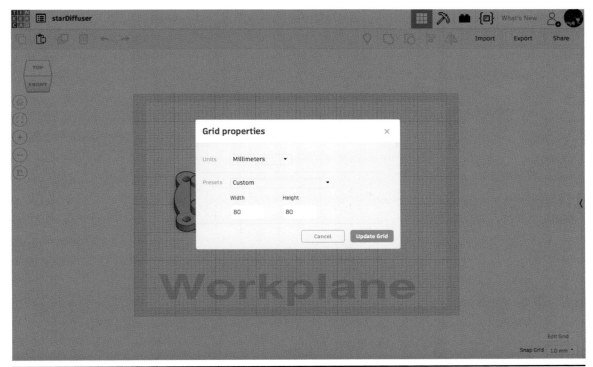

Figure 2.14   Edit "Grid properties" to resize your work plane to match the size of your printer or pattern piece.

Tinkercad gives each new design a whimsical name; you can change the name by clicking on the name in the upper right-hand corner.

Slicing with Cura

Once you are satisfied with your design, it's time to prepare the design for printing. First, export the design as an .stl file. In Tinkercad, the Export button is on the right-hand side of the toolbar. In order to print the design, you will have to translate the geometry of the model into commands that the printer can understand. This is usually referred to as *slicing*. 3D printers lay down a single 2D layer of melted plastic at a time. Adding additional layers on top of the first builds up the design in three dimensions. The layers are typically in the 0.06- to 0.15-millimeter range depending on the capabilities of the printer.

Slicing is the automated process of breaking a model into a series of thin layers and creating the machine control instructions for each layer. There are a number of 3D printer slicing apps. For these projects, we will be using Cura 3.3, a free and open-source slicer that works with many 3D printers (Figure 2.15).

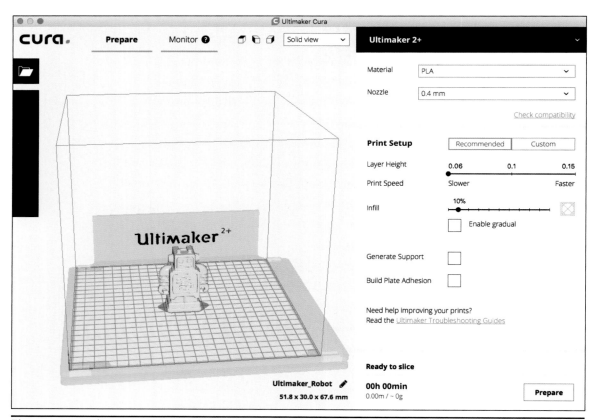

Figure 2.15   Slice your 3D models with Cura for 3D printing.

Downloadable Files for This Book's Projects

The projects in this book have sewing patterns, laser-cutter designs, 3D design files, and program files available as downloads from the publisher's website. Sewing patterns have a file extension of .pdf and can be printed on regular $8\frac{1}{2}$- × 11-inch printer paper. Designs that can be cut and etched on a laser cutter have a file extension of .svg. These can be read with most laser-cutter software as well as 2D design software. The 3D design files have an extension of .stl. 3D design models are also available on Thingiverse (www.thingiverse.com/amped_atelier/designs) and YouMagine (www.youmagine.com/amped_atelier/designs), or you can have the models printed by Shapeways (www.shapeways.com/designer/amped_atelier/creations).

CHAPTER 3

3D Embellished T-Shirt

USE **3D** PRINTING ON FABRIC TO MAKE UNIQUE CLOTHING AND COSPLAY ACCESSORIES.

What You Will Learn

In this chapter, you will learn how to

- Create a 3D design for printing on fabric
- 3D print on fabric

Files You Will Need

- Small adult unisex T-shirt pattern: pattern_small.pdf
- Medium adult unisex T-shirt pattern: pattern_medium.pdf
- Large adult unisex T-shirt pattern: pattern_large.pdf

Tools You Will Need

- Tinkercad or other 3D modeling tool
- Cura or other 3D slicing software
- 3D printer
- Sewing machine and sewing supplies

Materials You Will Need

- One yard of stretch knit fabric for T-shirt body (jersey, interlock knit, or similar)
- One yard of net, lace, or any fabric with small regular holes for T-shirt sleeves, two-way stretch recommended
- Small plastic clips to attach fabric to the 3D printer build platform (Wonder Clips or similar work well [www.clover-usa.com/en/sewing-and-quilting/105-wonder-clips-50 -pieces.html]).
- Polylactic acid (PLA) filament is recommended; nylon, acrylonitrile butadiene styrene (ABS), or other filament may be used.
- All-purpose polyester thread
- Ballpoint or stretch sewing machine needle
- No-sew option: Purchased T-shirt or other clothing to apply 3D-printed decoration

This raglan sleeve T-shirt has 3D-printed decorative elements printed directly on the fabric. You can 3D print directly onto fabric with your home 3D printer using standard PLA, ABS, or other regular filament. 3D printing on fabric retains the flexible properties of the fabric, does not require exotic filament, and permanently incorporates the 3D elements into the garment. The sleeves are made from a material with many small holes, so select net, lace, or any similar material. A two-way stretch fabric, meaning it stretches left-right but not up-down, is recommended for the sleeves, but non-stretch fabrics can be used as well. The fabric is sandwiched between printed layers of the design. The holes in the fabric allow the layers on top of the fabric to bond with the layer beneath it. Thick fabrics or anything with a large surface decoration are not recommended because these will interfere with the extruder (Figure 3.1).

To make this shirt you can download the pattern from the book web page, or for additional sizes and variety, you can use a similar T-shirt pattern purchased from a store or an online retailer. The raglan T-shirt pattern pdf with the book materials comes in adult unisex sizes small, medium, and large. Use the size of a well-fitting T-shirt to decide which pattern size to use for this project. Be sure to print out the included pattern at 100 percent size. Use a ruler to measure the 1-inch box on the printout to ensure the pattern is printing at 100 percent scale. If you are printing your pattern on standard copier paper, tape the 8½ × 11 inch sheets of paper together before cutting out the pieces of the shirt. Commercial raglan T-shirt patterns are available at fabric stores, craft stores, and on the Internet. Simplicity Patterns (www.simplicity.com) makes several easy to sew versions including Burda Style pattern number B9346 Child's Raglan Tops, Simplicity Child and Adult unisex pattern number 8223, Simplicity Men's Knit Top pattern number 8613, and Simplicity

Misses' Knit Top pattern number 8423. Kwik Sew pattern number K4146 (https://kwiksew .mccall.com/k4146) is a raglan shirt pattern in women's sizes 1X to 4X. Women's raglan sleeve shirt patterns up to plus size 3XL can be purchased from online retailers. Melly Sews Rivage Raglan (mellysews.com) and the Greenstyle Creations Centerfield Raglan (greenstylecreations.com) are both available in women's plus sizes.

Figure 3.1   Fabrics to use with 3D printing include net, tulle, lace, or other similar fabric with holes.

Step 1: Cut Out the Shirt Pieces

1. Use the raglan shirt PDF pattern from the book website or any raglan sleeve shirt pattern. The PDF pattern comes in unisex small, medium, and large sizes with T-shirt-length sleeves. Cut the front, back, and neck pattern pieces from stretch knit fabric. Jersey or interlock knits are typical T-shirt fabrics and are good choices for this project. The front and back pieces are each cut on the fabric fold so that they unfold to form the full fabric piece with no front or back seam.

2. Cut the left and right sleeves from net, tulle, lace, or other such fabric. You will 3D print decorative elements on the sleeve piece before assembling the shirt.

Step 2: Create Your 3D Design

When designing for 3D printing on fabric, keep individual shapes to no larger than 4 inches square. You can print names or words with spaces between the letters. Arrays of geometric shapes are good choices. Large designs with no spaces between the elements will not be flexible and should not be used as designs for this project. Once you are familiar with the basics of 3D printing on fabric and assembling the T-shirt, more complex designs can be found online or modeled in CAD programs. The dragon scales example pictured at the start of this chapter was designed by David Shorey, and is available on Thingiverse (www.thingiverse.com/thing:2755451).

1. Set the work plane in Tinkercad to the size of your printer bed.

2. Use Tinkercad's basic shape generator to lay out shapes of various sizes and heights. Larger, taller, and a greater number of shapes in your design will take longer to print. A design of between three and ten basic shapes that are approximately 0.5 to 1.5 inches each in length, width, and height is a good size to use (Figure 3.2). You can also search Thingiverse for designs and import those into Tinkercad as a base for your design.

3. To be sure that your design fits on the sleeve pattern, export the design as a SVG (Scalable Vector Graphics) file, and then print it on regular home paper printer. Lay the paper printout under or over the pattern piece to get an idea of how your design will look.

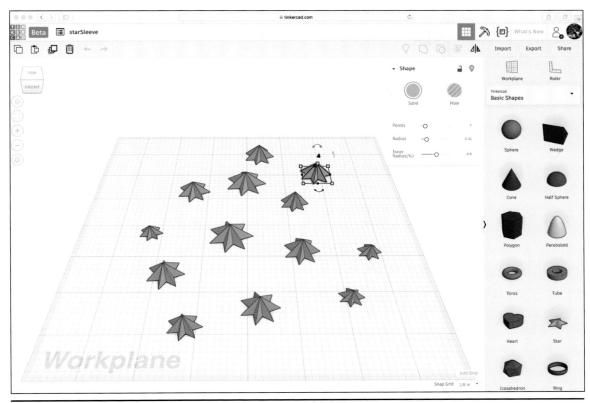

Figure 3.2   Stars created using Tinkercad's basic star shape generator.

Step 3: Export and Slice the Model

1. Export your design from Tinkercad as an STL file.

2. Import the downloaded model into Cura or other slicing software. Use the standard settings, and set infill to between 15 and 30 percent. Your design should not need supports, so be sure that this function is disabled. Your design also should not need the Cura setting for build plate adhesion (Figure 3.3).

3. Because the fabric is sandwiched between a couple of layers of the print, you will need to pause the print while it is printing so you can lay the fabric on the first layers.

    a. You can pause the print manually on some printers by waiting for a couple of layers to print and then pressing the Pause button; other printers might not have a Pause button, so you will need to put a Pause command in the print file.

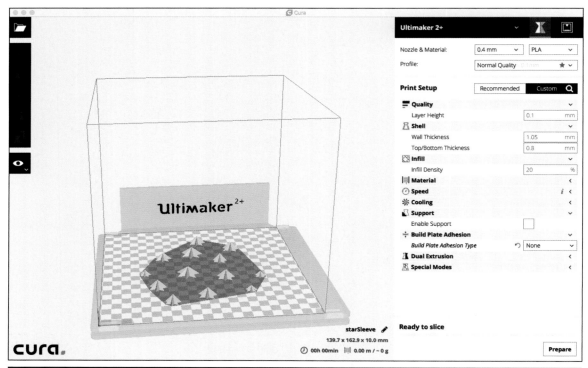

Figure 3.3   Star design Tinkercad model imported into Cura.

b. Cura has a plugin to pause the print execution. Select "Modify G-Code" in the Post Processing submenu under the Extensions menu. This should bring up the Post Processing Plugin window.

c. Add the "Pause at Height" script. Change to Pause at "Layer No." and set the layer number to 2. The printer will pause after the second layer has printed. Make sure that the "Resume Temperature" is set to the preferred printing temperature for your material (Figure 3.4).

4. Once you have selected your settings and added the optional Pause command, you are ready to slice. Press the Prepare button in the lower right-hand corner, and save the resulting G-code file. Transfer the G-code file to your printer according to your printer instructions using a web interface, SD card, or other method.

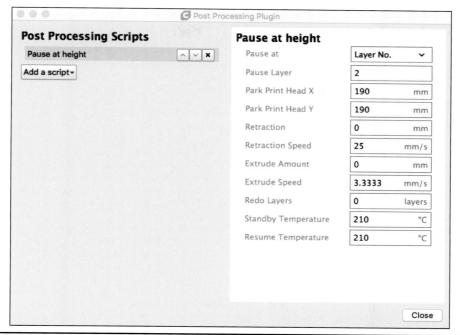

Figure 3.4 Use Cura's Post Processing Plugin to pause the printer after laying down a couple of layers.

Step 4: 3D Print on Fabric

1. Set up your printer for your 3D model and filament type. Print two to three layers of your design. Pause the printer using the manual printer controls or the Pause Print command in the G-code (Figure 3.5).

   Caution: The print nozzle is still extremely hot at about 180 degrees Celsius (356 degrees Fahrenheit). Do not touch it! If you are worried about touching it, wear some work gloves during this step.

2. Carefully lay one sleeve fabric piece fabric down over your 3D-printed layers. An advantage of this printing method is that you can precisely place the fabric because you know exactly where the design is printing. Smooth the fabric on the build plate, and use numerous clips to hold the fabric taut at the middle and corners. Check to be sure that your fabric won't get caught in any of the moving parts of the printer and that the extruder head won't hit any of the clips while it is moving around to print (Figure 3.6).

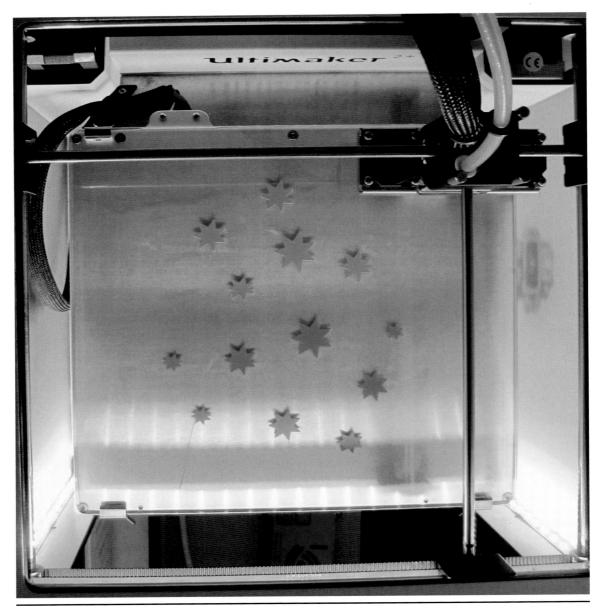

Figure 3.5   Pause the printer after a couple of layers have printed.

3.   Resume printing. Carefully watch the printer for the next couple of layers to be sure that the extruder head is not pulling the fabric too much. A couple of small wrinkles are generally not a problem, but if your fabric is too loose, the extruder may drag the prints off center. The printer is laying down minuscule amounts of filament that are often hard to see on the first layer on the fabric. When the print is finished, wait for the build plate and nozzle to cool, and carefully remove your fabric and 3D-printed design.

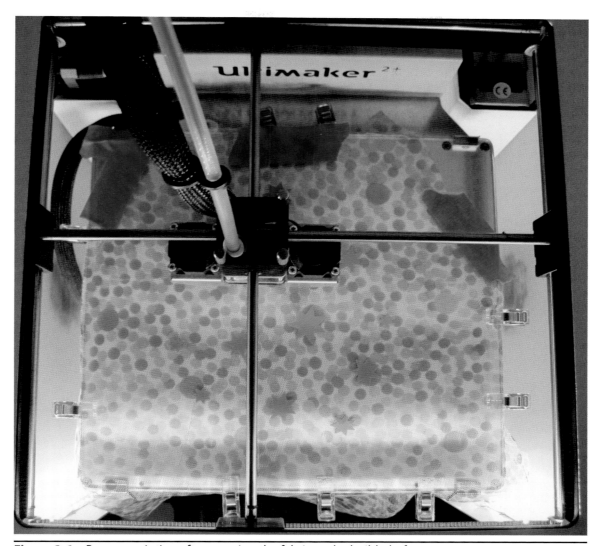

Figure 3.6   Resume printing after securing the fabric to the build platform.

4. If you have a symmetrical pattern, repeat step 4 with your second sleeve piece. You can also mirror an asymmetrical pattern to print matching right and left sleeves. Cura provides several tools to adjust your model before slicing. The Mirror tool is located on the toolbar below the Rotate tool. Clicking on the Mirror tools brings up six arrowheads, two for each axis. To mirror the model, simply click on one of the arrowheads. To mirror the shapes, left to right click the red arrows (Figure 3.7).

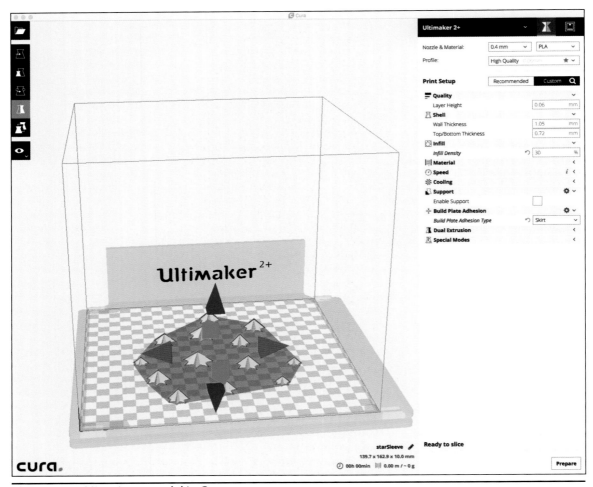

Figure 3.7   Mirroring a model in Cura.

Step 5: Assemble the T-Shirt

The PDF raglan sleeve T-shirt pattern pieces have a $\frac{1}{2}$-inch seam allowance. Use a zigzag stitch to sew all the seams to keep the stretch in your fabric. A serger can also be used to sew the shirt together. In the instructions that follow, "right side" of the shirt describes the fabric that will be facing out when the shirt is worn, and "wrong side" is the side that is toward the body when the shirt is worn.

1. Place the front side of the sleeve over the shirt front with the right sides facing together. Using a zigzag stitch, sew the front side of the sleeve to the shirt front. Repeat for the other side (Figure 3.8).

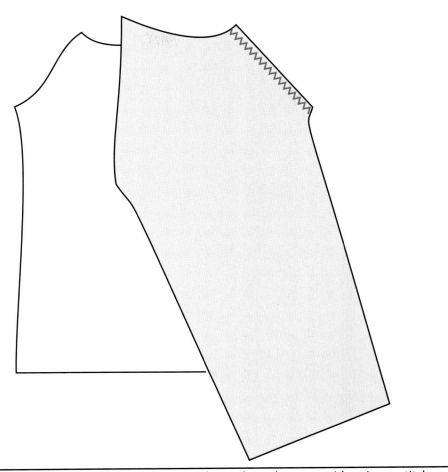

Figure 3.8 Attach the sleeve to the front piece, and sew along the seam with a zigzag stitch.

2. Place the back side of the sleeve over the shirt back with the right sides facing together. Sew the back piece to the sleeve. Repeat for the other side. You should now have both sleeves sewn to the front and back. Lay your shirt out, right side facing up.

3. Sew your neckband with the right sides together at the short ends to form a loop. Fold the neckband in half along the long edge with the wrong sides facing in and the raw edges of the fabric at the bottom. Using pins as markers, divide the neckband into four equal lengths: first, place a pin directly opposite the seam. Then align the seam and the pin, and place a pin at each of the folds.

4. Divide the neckline into quarters by aligning the front and back centers and putting a pin halfway between the two (it will be on the sleeve toward the front of the shirt). Align the seam in the neckband with the center back. Align the other three pins accordingly. Continue pinning the neckband around the neckline, stretching evenly to fit. The neckband is smaller than the collar, so it will not fit exactly (Figure 3.9).

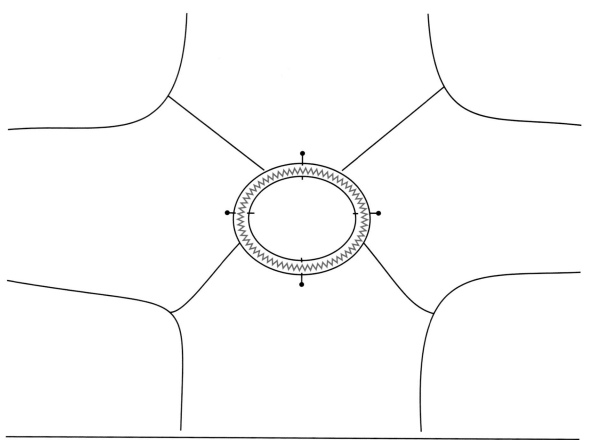

Figure 3.9   Pin the neckband to the collar in quarters, and sew the neckband to the collar.

5.   Sew the neckband to the neckline, stretching the neckband between pins. Flip the neckband up, and press the seam allowances down toward the shirt. If desired, top stitch around the neckline, catching the seam allowances underneath in the inside of the shirt.

6.   Turn the shirt inside out. Line up the sleeves and side seams at the raw edges. Starting at the wrist, sew the sleeves together up to the underarm seam, and continue sewing down the side seam in one sewing line. Repeat for the other side (Figure 3.10).

7.   Turn 1 inch on the hems on the sleeves and the bottom hem to the inside, and sew with a zigzag stitch.

No-Sew and Low-Sew Options

You can use the techniques in this chapter to make a 3D-printed on fabric patch that can be glued using fabric glue or sewn onto a premade clothing item, such as a T-shirt, jacket, or hoodie (Figure 3.11).

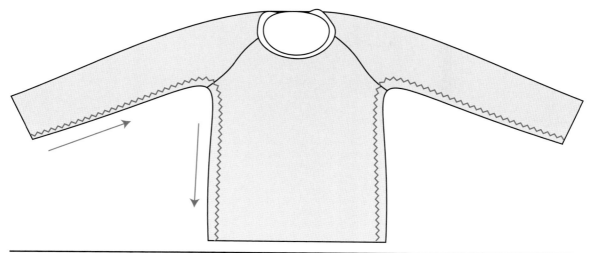

Figure 3.10 Sew the side seams of the shirt from wrist to waist.

Figure 3.11 3D-printed pink and purple shirt. (*Courtesy of Jason Martineau*)

CHAPTER 4

Fiber-Optic Fabric Scarf

MAKE AN ETHEREAL GLOWING LIGHT-UP SCARF USING FIBER-OPTIC FABRIC AND A SUPER-BRIGHT LED. This project looks great in low-light situations, such as outdoors at night or in dark indoor locations.

What You Will Learn

In this chapter, you will learn how to

- Work with fiber-optic fabric
- Add a light source using a simple electric circuit
- Sew bias tape on a raw fabric edge

Files You Will Need

- Laser cutter–ready scarf pattern: scarfLong.svg and scarfShort.svg
- 3D model: ledBatteryHolder.stl

Tools You Will Need

- 3D printer
- Sharp fabric scissors
- Laser cutter or hobby knife
- Sewing machine, thread, and sewing tools

Materials You Will Need

- $1\frac{1}{2}$ to 2 yards of fabric that can be cut with a laser cutter: cottons, cotton-poly blends, tightly woven linens or silks, wool felt, and natural or synthetic velvets. Do *not* use fabrics with polyvinyl chloride (PVC), including vinyl and some synthetic leathers, which can release toxic fumes.

- 4 yards of double-fold or extra wide double-fold bias tape

- Fiber-optic panel/fabric (www.sparkfun.com/products/12712)

- Black electrical tape

- 5-millimeter super-bright light-emitting diodes (LEDs; www.sparkfun.com/products/531): white, green, and red recommended

- 2 coin-cell 3-volt (V) batteries

- Opaque 3D printer filament or printed 3D battery holder

No-Laser Option

This project can be made without the laser-cut fabric by replacing it with purchased fabric with large holes or a chunky lace such as guipure lace.

Figure 4.1   Materials for the fiber-optic scarf.

An optical fiber is a strand of glass or plastic designed to carry light from one end to the other through internal reflection. Fiber-optic fabric has strands of plastic optical fiber woven as the warp so that the length of the fabric lights up. The optical fibers were sanded and nicked along their length so that light leaks out of the length of the fibers for a sparkly effect.

In the fiber-optic fabric we use, purchased through Sparkfun, the optical fibers have been gathered into a bundle similar to the size of a basic LED. Holding a powered LED up to the bundle will light the entire fabric.

To see how it works, you can light up the fiber-optic fabric using a basic LED and a coin-cell battery. Because a 3-volt (V) coin-cell battery can't source enough current to damage the LED, we can connect the LED directly to the battery. Slide a 3-volt (V) coin-cell battery between the legs of the LED with the long leg (+) on the top of the battery (+) where the writing is located and the short leg on the bottom. Pinching the legs should light up the LED. Place the top of the LED up against the bundled end of the fabric, and the fabric should light up. Remember that it works best in low light (Figure 4.2).

Figure 4.2   Fiber-optic fabric lit by holding an LED to the bundled end of fibers.

We'll prepare the fiber-optic fabric, create a simple light source, laser cut the scarf pattern, and put it all together.

Step 1: Laser Cut the Fabric for the Scarf

1. Download the laser files for the fiber-optic scarf. The file sizes are set for a laser bed size of 18 inches by 24 inches. Depending on the bed size of your laser cutter, you may need to resize these files in a CAD program such as Inkscape or Adobe Illustrator. Each fabric panel is 40 × 75 centimeters (approximately 16 × 30 inches).

2. Cut four scarf pieces and two connector pieces of the fabric with the laser cutter. These will be enough for the front and back of the scarf.

Step 2: Split the Fiber-Optic Panel

Our panel is 40 × 75 centimeters, or about 16 inches wide by 30 inches long. If your panel has different dimensions, adjust the project accordingly. The panel is a little short for a scarf, so we'll cut it into two narrow pieces to make a longer scarf. Since optical fibers reflect light internally down their length, cutting across the fibers would disable or cut off the light. In order to make two useful pieces, we will cut lengthwise between the optical fibers.

1. Measure along the 40-centimeter width of the fabric, and at 20 centimeters, draw a chalk or pencil line up the 75 centimeters in length of the fiber-optic panel toward where the strands are bundled.

2. Sew two parallel lines of zigzag stitching along the length on either side of the line. These stitches will help to keep your fiber-optic panel fabric from unraveling. Remember to backtack at the beginning and ends of your stitching (Figure 4.3).

3. Cut along the chalk or pencil line, trying to cut as few fiber-optic fibers as possible. As you approach the bundle of fibers, you will need to carefully separate the fibers to avoid cutting them. At the top, above the fiber-optic panel, continue to cut the fabric. You will have to carefully cut above and below the fiber bundle. Leave the bundled fibers at the top intact; just cut the fabric (Figure 4.4).

Figure 4.3 Sew parallel lines of zigzag stitching along the center length of the fiber-optic panel.

4. Sew the panel pieces together by hand or machine using a thread color that matches your laser-cut fabric. To create one long scarf piece, you will need to sew the ends of the panel fabric together with the bundle of fiber optics in the center. Overlap the panel pieces approximately 4 inches. Sew the panel fabric together along the overlap using a zigzag stitch, being careful not to hit the fiber-optic bundle with your sewing machine needle. You may have to make a small slit in both tops of the fabric to allow the fiber-optic bundle end to be repositioned so that the fabric lies flat. You can also sew this together by hand with a running stitch (Figure 4.5).

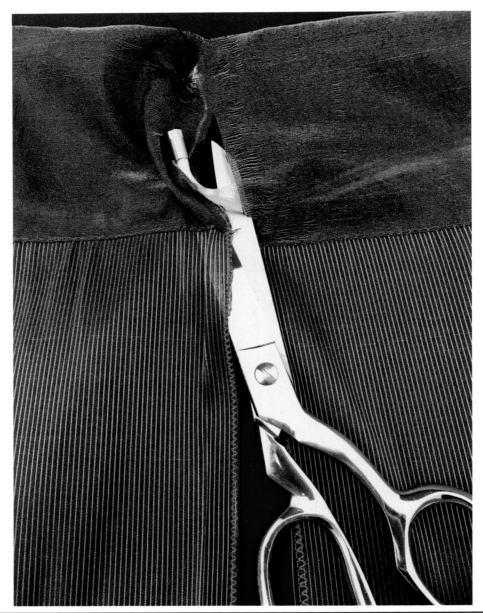

Figure 4.4   Split the fabric by carefully guiding your scissors between the warp optical fibers of the panel. When you get to the top, just cut the fabric, not the fiber optics.

Figure 4.5 Overlap the ends of the fiber-optic panel, and sew along the overlap to make one long scarf.

Step 3: Sew the Scarf Fabric and Attach It to the Fiber-Optic Panel

1. Sew three fabric pieces together on the short-width end with a ½-inch seam allowance to make one long piece. Repeat for the other three pieces. Press the seams flat.

2. Place one long fabric piece on each side of the fiber-optic panel with the wrong side of the fabric facing the fiber-optic panel. Pin along the long sides of the scarf. Leave space for the metal fiber-optic bundle holder to stick out of the fabric. We'll need to attach the LED to this in a later step.

3. Sew along both long sides using a regular machine or running stitch, removing the pins as you sew. If your fabric ends up being too long for the fiber-optic panel, you can trim the end of the fabric using scissors. If the long ends are uneven, trim the fabric so that there is an even edge on each side of the scarf (Figure 4.6).

Figure 4.6   Sew the fiber-optic panel to both sides of the cover fabric along the long edges, and then trim to even out the fabric.

4. You can leave the scarf fabric as is, or you can apply bias tape to enclose the long edges of the scarf for a more finished appearance. With any bias tape, the back side of the tape is slightly wider than the front, so when you stitch the tape on from the front, it will be easier to catch the wider edge on the back side of the project.

5. Measure a piece of bias tape that matches the long side of your scarf plus 1 inch for folding the ends under. Cut two pieces this length.

6. Completely open up the three folds of the bias tape fold to find the wider side of the tape. Fold the end you cut of the bias tape in $\frac{1}{2}$ inch so that you have a nice folded edge at the bottom of the scarf. With the tape open, pin the wide side of the tape to the back side of the project, aligning the raw edges (Figure 4.7).

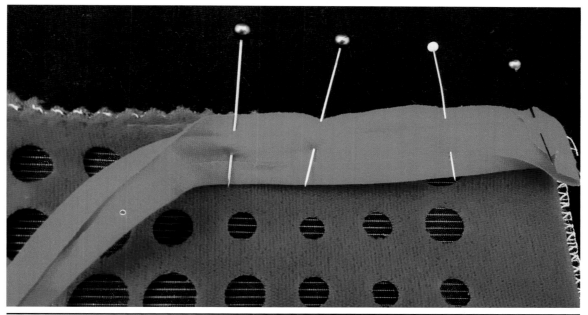

Figure 4.7 Fold the bias tape end over at the bottom for a neat start, and pin the wide side of the tape to the back side, aligning the raw edges.

7. Stitch along the folded crease of the tape, removing pins as you go. Be sure to backtack at the beginning and end of the stitching.

8. Place the project face up on the work surface. Wrap the bias tape up and over the raw edge of the fabric, and secure it with pins. The edge of the bias tape should just cover the line of stitching from the preceding step. The tape should lie flat along the fabric edge (Figure 4.8).

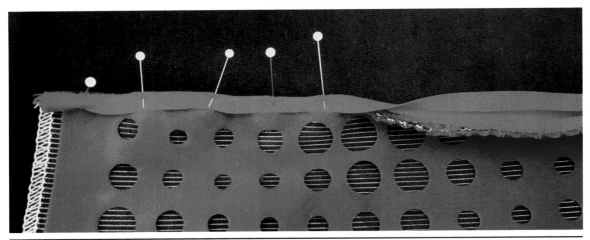

Figure 4.8   Wrap the bias tape over the edges of the fabric so that it just covers the line of stitching.

9.   Stitch as closely as possibly to the front edge of the bias tape, removing the pins as you stitch and backtacking at the beginning and end of the stitching. Your stitching should catch the back side of the bias tape to completely encase the edges of the scarf (Figure 4.9).

Figure 4.9   Topstitch the edge of the bias tape to completely encase the edges of the scarf.

Step 4: 3D Print the Battery and LED Holder

Print the case that holds the fiber-optic bundle, LED, and battery. Printing with an opaque filament will block light from leaking out of the LED connection (Figure 4.10).

1. Download the 3D model of the led battery holder: ledBatteryHolder.stl. 3D print with these print settings: 10 percent infill, no supports, and a brim for build-plate adhesion.

2. The model should be able to print without support if it is standing with the tube at the bottom.

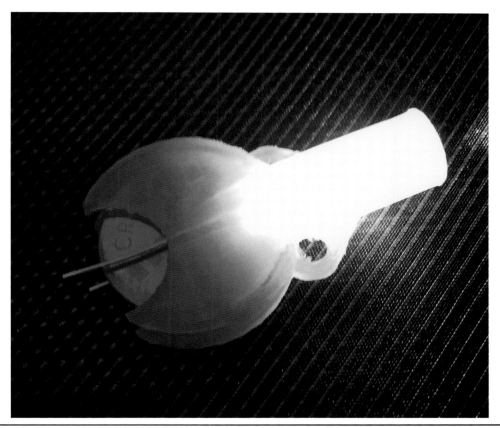

Figure 4.10   3D-printed battery holder with a lit LED and coin-cell battery.

Step 5: Attach the Light Source

The LED and the fiber-optic bundle being about the same width fit in the tube end of the battery holder. Once assembled, they can be held together with electrical tape.

1. Find the metal fiber-optic bundle holder in the center of your scarf. Carefully pull the bundle through the fiber-optic panel and your scarf material. If it doesn't fit through

the holes in the laser-cut material, you may need to cut a small slash with sharp scissors. Pull your sewn scarf to one side so you have full access to the bundle (Figure 4.11).

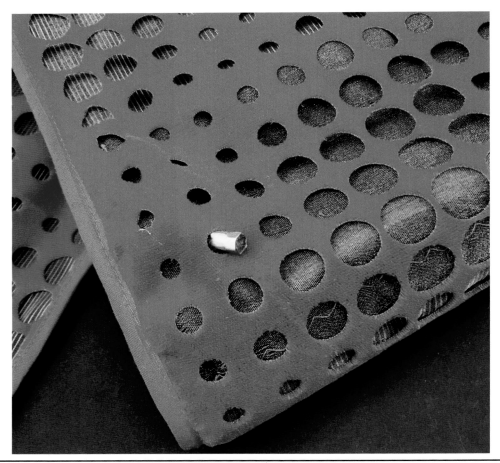

Figure 4.11   Pull the bundle connector through a hole in the scarf.

2. Insert the LED with the wire leads first through the tube end of the battery holder until the LED does not go any farther. Then insert the metal end of the fiber-optic bundle holder into the tube so that it fits right up to the dome part of the LED. Tightly tape the tube end of the battery holder and the fiber bundle with electrical tape so that they don't come apart (Figure 4.12).

3. Place a battery between the leads of the LED with the positive side of the battery touching the long LED leg. The holder should keep the battery in place through tension.

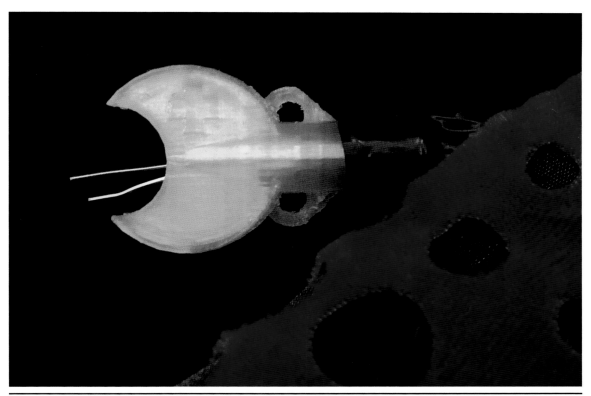

Figure 4.12 Tightly tape the fiber-optic bundle to the LED in the battery holder with electrical tape.

4. Whipstitch the battery holder to the outer fabric through the sewing holes on the sides. You can tuck the LED and battery inside the outer fabric before you sew on the battery holder, but be aware that this will make it harder to access the battery.

5. This project does not have an on/off switch. As long as the battery is held correctly between the legs of the LED, the scarf will illuminate. To turn the scarf off remove the battery from the holder.

Going Further

This project introduces a simple circuit that lights up an LED when it is in contact with a battery. A next step for this project would be to add a switch to the circuit so that you can turn the scarf on and off without having to take out the battery. The circuit can be connected with wire or with conductive thread. Take this project further by adding a simple light sensor, such as a photoresistor or photocell, that can detect light so that the scarf only lights up in the dark.

CHAPTER 5

Fun Festival Hip Pack

LET'S START PROGRAMMING WITH THE ADAFRUIT CIRCUIT PLAYGROUND EXPRESS MICROCONTROLLER board and Microsoft's browser-based visual programming environment MakeCode. we'll also 3D print a decorative cover for the board, sew the hip pack in your choice of fabrics, and put it all together so that your pack reacts to music and sound.

What You Will Learn

In this chapter, you will learn how to

- Create a sound-reactive light-up festival hip pack using visual block programming

Files You Will Need

- Hip pack pattern: hipPackPattern.svg
- Circuit Playground cover 3D model: cpxCover.stl
- MakeCode program: cpxFestival.uf2

Tools You Will Need

- Sewing machine and sewing tools
- Computer with browser
- USB Micro B cable
- 3D printer (optional)

Materials You Will Need

- $\frac{1}{4}$ yard of woven fashion fabric (Novelty prints or metallics are recommended.)
- $\frac{1}{4}$ yard of woven lining fabric
- 12-inch sew-on hook and loop fastener tape (Velcro)
- $1\frac{1}{2}$ yards of 1 inch wide belting or webbing for belt (This is a long, adjustable length for an average adult, you may need more or less depending on your waist measurement.)
- 1-inch parachute buckle
- Adafruit Circuit Playground Express (Product ID 3333, www.adafruit.com)
- AAA battery pack (Product ID 727, www.adafruit.com)
- AAA batteries
- For 3D-printed cover: Polylactic acid (PLA) filament in natural or light translucent colors or transparent flexible filaments
- No-sew option: Purchased hip or waist pack

The Circuit Playground Express is a great learning board with 10 built-in addressable light-emitting diodes (LEDs) and packed with sensors. It is easy to program in three different ways, and extra lights can be added by sewing them on with conductive thread or soldering them to the board with wire. MakeCode is a web-based editor with colored blocks that you can drag and drop into the workspace to construct a program. An interactive simulator provides instant feedback on how the program is running without having to run the program on a microcontroller board. You can learn how to use MakeCode through the tutorials and projects at https://makecode.com/.

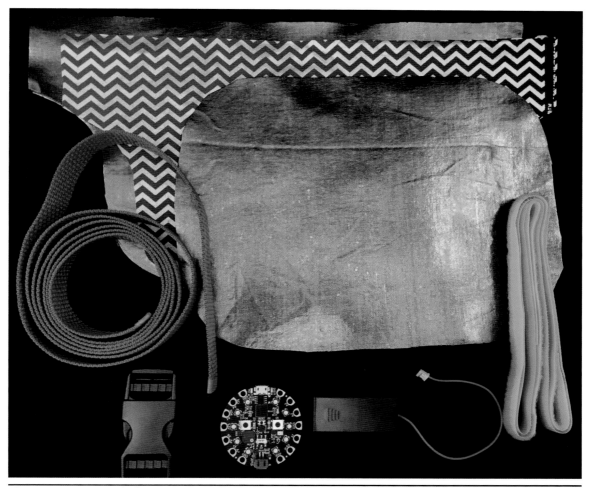

Figure 5.1   Materials for constructing the hip pack and the Circuit Playground Express board.

Step 1: Program the Circuit Playground Express

Adafruit has developed extensive guides and tutorials for the Circuit Playground Express. If your board isn't working as expected or you want to learn more, start with the guide at https://learn.adafruit.com/adafruit-circuit-playground-express/overview.

1. To get you started, we've made a sound-activated light-up program for you to use with your Circuit Playground Express. Download the cpxFestival.uf2 file from the book's website to use in this project, or make your own in MakeCode (https://makecode.com) and save it as a .uf2 file (Figure 5.2).

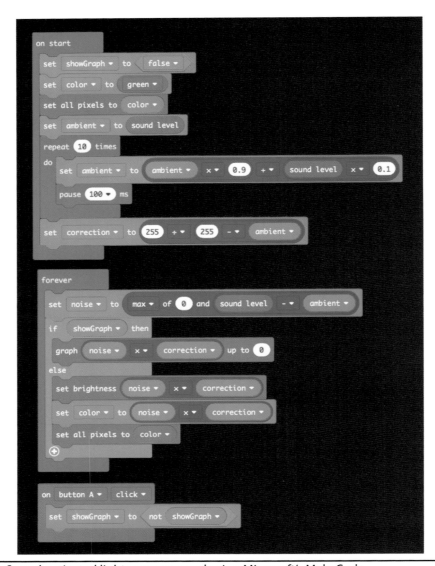

Figure 5.2   Sound-activated lights programmed using Microsoft's MakeCode.

2.   Before you plug in your board, you might have to install a driver. If you have a computer running Windows, you will need a driver; if you are running a Mac or Linux, you will not need to install a driver. To install the driver, go to https://learn.adafruit.com/ adafruit-circuit-playground-express/adafruit2-windows-driver-installation. Download and run the installer.

3.   Connect your computer with the Circuit Playground Express using a data sync USB Micro B cable. Some USB cables only charge devices, so try a different cable if you are not getting any results.

4.   Press the Reset button once to enter Bootloader mode. You may need to double-press Reset if this is your first time using MakeCode with your Circuit Playground Express.

5. All the LEDs should briefly turn red and then turn green, and the status LED should pulse red. Your computer will have a new removable drive called CPLAYBOOT.

6. Find the cpxFestival.uf2 file and copy/paste or drag and drop it to the CPLAYBOOT drive.

7. The status LED on the board will blink while the file is transferring. Once it's done transferring your file, the board will automatically reset and start running your code.

8. You can now unplug your Circuit Playground Express from the computer, and it will keep running the program.

Step 2: 3D Print the Circuit Playground Express Cover

A 3D-printed decorative cover will take your project to the next level and beautifully diffuse the LEDs on the Circuit Playground Express. Natural-colored or light-colored translucent PLA, such as yellow or amber, is recommended for the LEDs to look best. Translucent flexible filaments or glow-in-the-dark PLA also may make a great cover.

1. Download the model file: cpxCover.stl.

2. Load the model into Cura for slicing, and use these print settings: 10 percent infill, with supports, and no build adhesion.

3. Load the translucent filament, and start your 3D printer according to the operating instructions. Print!

Step 3: Make the Pack

As you sew the bag, remember to backtack at the start and stop of each seam to lock the stitches in place. If you make a mistake, use your seam ripper to undo the stitches and try again.

1. Cut Velcro into two 6-inch-piece sets. Then, using a straight stitch or zigzag stitch, sew the hook and loop tape to the pattern pieces, where they are marked: (1) hook side below the fold line in the pocket piece, (2) loop side in a matching position on the lower half on the right side of the front piece, (3) hook side at the top of the right side on one lining piece, and (4) loop side at the top of the matching top of the right side of the other lining piece (Figure 5.3).

Figure 5.3   Sew the hook and loop tape on the lining pieces, front section, and pocket.

2. Fold the pocket piece in half, wrong sides together with the fold at the top and the corners at the bottom, and press.

3. Line up the pocket piece with the right side of the front piece, and secure them together using the sewn-on hook and loop tape (Figure 5.4). This is now the front section.

4. Cut belting into two pieces measuring 1 yard and ½ yard.

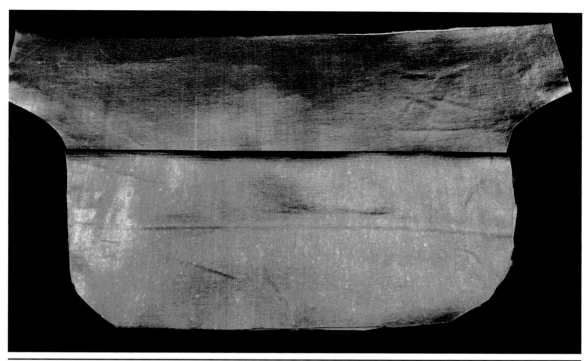

Figure 5.4 Pocket piece attached to the front with the hook and loop tape.

5. Line up one end of the belting to the right side of the back section, and pin it ½ inch from the top edge. Sew along the edge to attach the belting to the right side of the back section. Sew over this stitching again to help secure the belt even more (Figure 5.5). Pin the other half of the belting to the right side of the back section, and sew along edge. Then sew it again along the same stitching line for a more secure attachment.

6. Place the front and back sections together, and pin them along sides and bottom edge. The belt will be sandwiched inside, pull the belt parts through the open top edge so that they don't get caught in your sewing by mistake. Sew along the sides and bottom, removing pins as you sew (Figure 5.6). Turn the bag right side out and press with an iron.

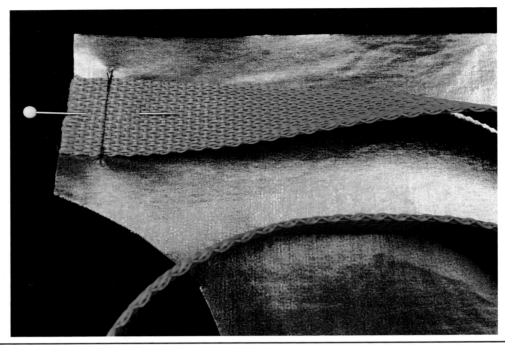

Figure 5.5   Pin and then securely sew the belting to the right side of the hip pack back section.

Figure 5.6   Sew the front and back sections together along the sides and bottom, keeping the belt out of the way.

7. Pin the lining section right sides together, and sew them along the sides. Sew the bottom edge keeping a 4-inch section open in order to turn the bag later (Figure 5.7).

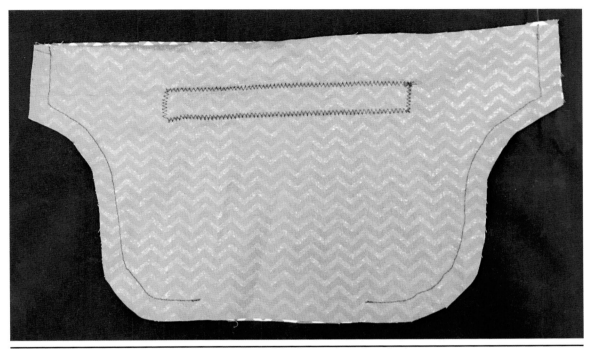

Figure 5.7   Sew along the sides and bottom of the lining, keeping the bottom center open.

8. Slip the lining over the right-side-out outer bag, and with the right sides together, pin the lining to the bag. The belt should be in between the lining and the bag.

9. Stitch all the way around the upper edge of the lining and bag (Figure 5.8).

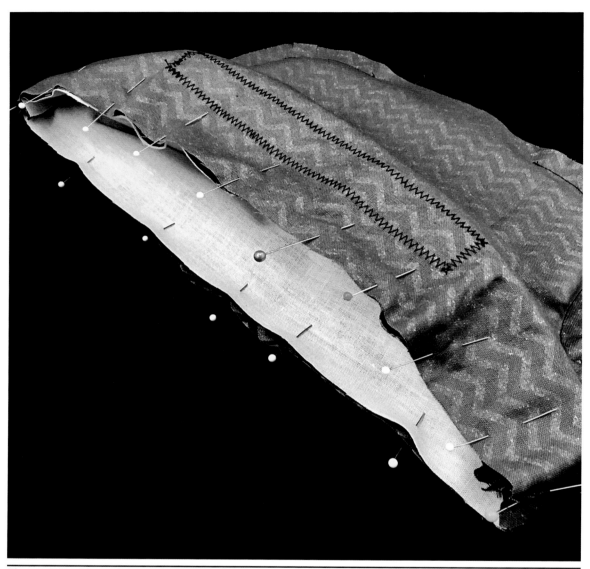

Figure 5.8   Pin and then stitch around the upper edge of the bag and lining.

10. Turn the lining right-side out by pulling the outer bag back down through the opening. Once the bag is out, turn the open edges at the bottom of the lining ½ inch inside, and whipstitch them together so that the lining seam is closed (Figure 5.9).

11. Put the lining inside the bag. The belt pieces should now be on the sides of the bag.

12. On one loose end of the belt, fold the belting over ¼ inch, then fold it again ¼ inch, and sew the folded edge down with a zigzag or a straight stitch to neaten the edges so that they don't unravel. Push the neatened edge of the belt through one side of the parachute buckle. Repeat this step with the other side of the belting and the second half of the parachute buckle (Figure 5.10).

Figure 5.9 After turning the bag out through opening, whipstitch the lining bottom closed.

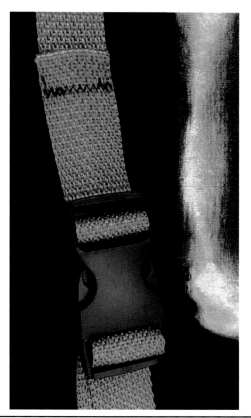

Figure 5.10 Insert each end of the belting into one side of the parachute buckle, then sew each end with a neat hem.

Step 4: Bring It All Together

1. Using regular sewing thread, whipstitch the Circuit Playground Express through four of the large connection pads to your pack. Use several stitches, but do not completely cover the copper connections of the pad (Figure 5.11). If you decide to build on your project, you want to leave some room for wire or conductive thread to connect.

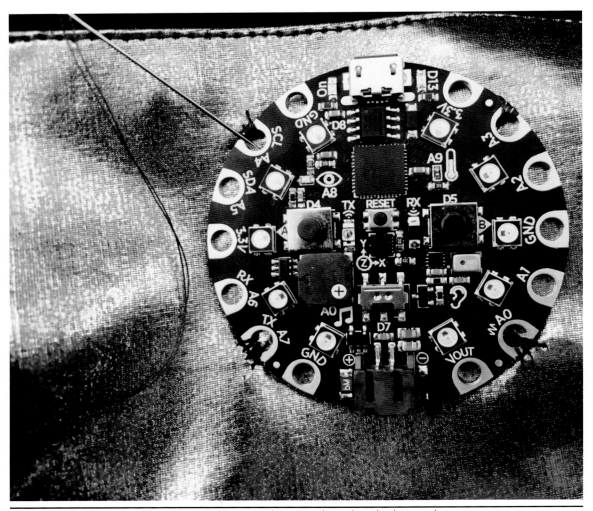

Figure 5.11   Whipstitch the Circuit Playground Express board to the hip pack.

2. Use regular sewing thread to whipstitch the 3D-printed decorative cover to the pack over the Circuit Playground Express.

3. Attach the battery pack, and tuck it inside the front pocket! Turn it on, and have some fun (Figure 5.12)!

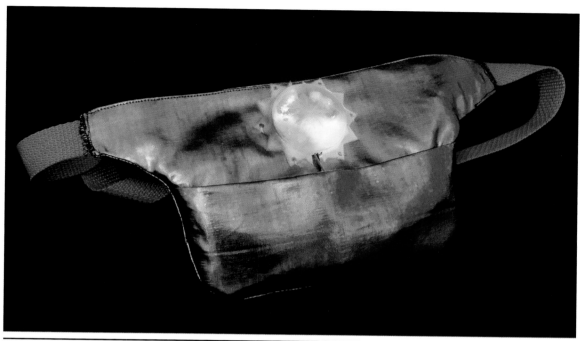

Figure 5.12   The hip pack ready for festival fun.

CHAPTER 6

Solar Backpack

HARNESS THE POWER OF THE SUN WITH THIS PERSONALIZED SOLAR BACKPACK THAT WILL KEEP your mobile devices charged.

What You Will Learn

In this chapter, you will learn

- The basics of photovoltaics
- How to put together a solar charger from existing subsystems
- How to construct a backpack

Files You Will Need

- Backpack pattern: backpackPattern.pdf

Tools You Will Need

- Sewing machine and thread
- Soldering tools: soldering iron, solder
- Basic electronics tools

Materials You Will Need

- Large 6-volt (V), 3.5-watt (W) solar panel (www.adafruit.com/product/500)

- USB/solar lithium ion/polymer battery charger (www.adafruit.com/product/390)

- Lithium ion/polymer battery: 3.7 volts (V), 1,200–2,500 milliampere hour (mAh)

- 3.5/1.3- or 3.8/1.1- to 5.5/2.1-millimeter DC jack adapter cable (www.adafruit.com/product/2788)

- Adafruit PowerBoost 500 Basic (www.adafruit.com/product/1944)

- High-temperature polyimide tape (www.adafruit.com/product/3057)

- 1 yard of 45-inch or $^3/_4$ yard of 60-inch outdoor fabric, canvas, denim, or other sturdy fabric

- $^1/_2$ yard of contrast-colored outdoor fabric, canvas, denim, or other sturdy fabric for top flap

- $^1/_4$ yard of lightweight iron-on interfacing

- 2 yards of 1-inch-wide webbing or belting

- 2 yards of $^1/_4$-inch cord or similar thin ribbon

- One 1-inch quick-release clip or parachute clip

- Four D-rings of 1-inch diameter (**Note:** The diameter of the D-rings should match the width of the straps.)

- **No-sew option:** Purchased backpack with a fabric or canvas top flap measuring at least 10 inches × 6 inches

Our four main electronics components in this project are the solar panel to convert light into electricity, the solar charger to regulate and manage charging a battery, a battery to store that electricity, and the power boost to charge our phone (Figure 6.1). The solar panel and solar charger for this project should only be used with 3.7-/4.2-volt (V) lithium ion/polymer battery. The solar charger features smart load sharing to automatically use the input power, when available, to keep the battery from constantly charging/discharging.

There are a couple of things to remember about working with solar chargers. First, don't leave the batteries in direct sunlight or in very hot places, like a car dashboard, because this will damage the battery and reduce the battery life! Second, solar panels work best in direct sunlight. Any covering, including glass, plastic, or shade, will decrease the efficiency of the panel. We'll be mounting our panel on the outside of the backpack to get the best access to the sun.

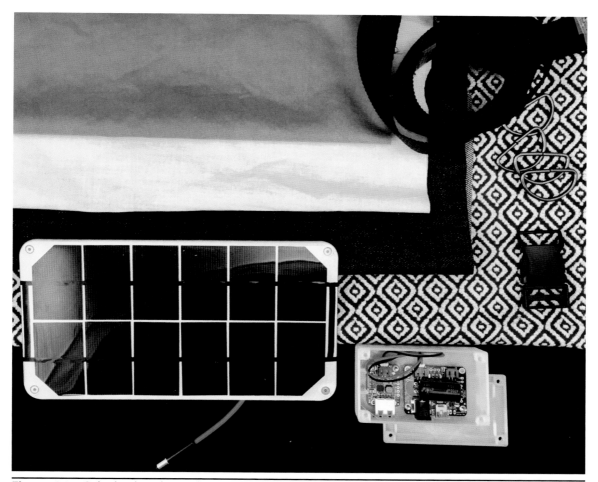

Figure 6.1 Solar backpack supplies.

Step 1: Prepare the Solar Charger

The Adafruit USB, DC & Solar Lipoly Charger comes assembled with all its components except for the large filtering electrolytic capacitor. Our first step is to solder the capacitor to the solar charger. Capacitors are polarized, so check the polarity of the capacitor to make sure that the positive lead of the capacitor goes into the pad marked with the plus sign (+). If you want, you can bend the capacitor over a bit as well, but don't have it touch the hot charging chip, the black square in the middle of the printed-circuit board (Figures 6.2 and 6.3). Cover the charging chip with high-temperature polyimide tape, sometimes referred to by its color and the brand name as yellow Kapton tape. Polyimide is both electrically isolating and can withstand high temperature.

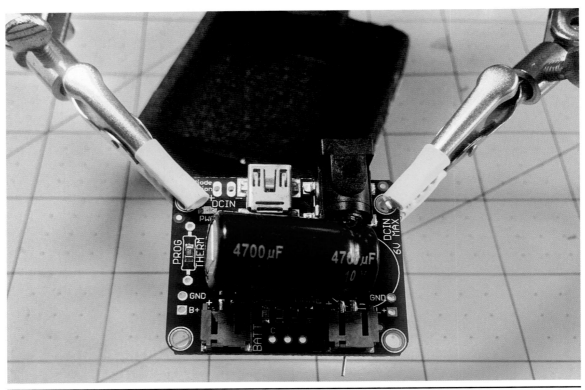

Figure 6.2 Mount the capacitor parallel to the charger board. Remember, the capacitor is polarized, so check the polarity.

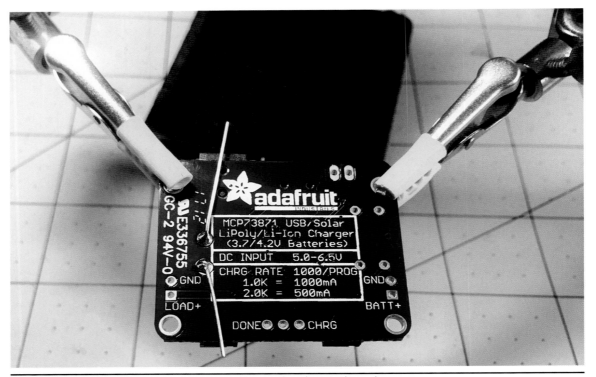

Figure 6.3 Solder the leads of the capacitor to the charger board.

Step 2: Prepare the Solar Panel

If the power connector of the solar panel doesn't fit into the DC jack of the solar charger, you will need to either swap out the DC plug or, if the solar panel came with a 1.1-millimeter plug, use the 3.5 to 3.8/1.1- to 5.5/2.1-millimeter DC jack adapter cable (www.adafruit.com/product/2788). The solar charger DC jack accepts a 5.5-millimeter-long and 2.1-millimeter-diameter plug. Let's go over swapping the plug.

1. Snip off the connector, and then strip off the outer casing (Figure 6.4).

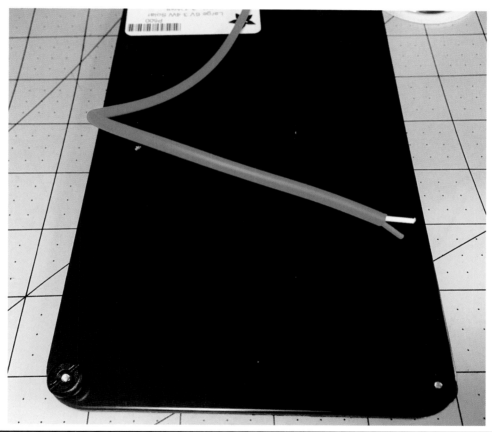

Figure 6.4 If you need to swap the solar panel power connector, snip off the connector and then strip off the outer casing.

2. Solder the 5.5/2.1-millimeter barrel connector (www.adafruit.com/product/3310) to the striped power cable. Strip the outer sleeve of the power connector to reveal the red and black wires. Strip and tin these wires, and then solder the red wire to the center connector and the black wire to the ring of the jack (Figure 6.5).

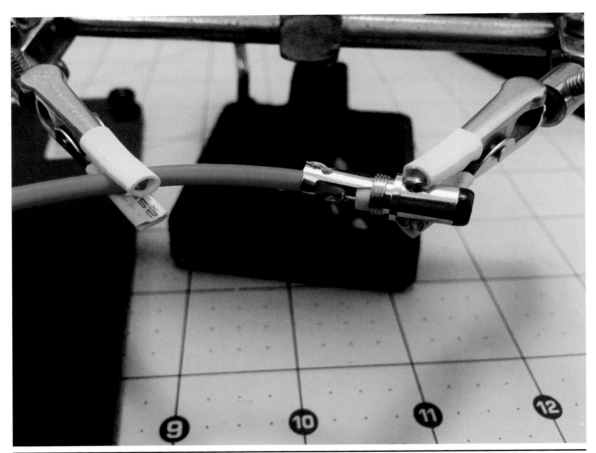

Figure 6.5   Solder the red wire to the center connector and the black wire to the ring of the jack.

Step 3: Solder the USB Connector on the Power Boost 500

The Power Boost 500 comes assembled, except for the USB plug. Solder the USB plug to the Power Boost 500 board.

Step 4: Splice Two JST-PH Connectors End to End

A short connector cable can be used to connect the solar charger to the Power Boost 500. Using a cable instead of soldering connections between components allows you to swap and replace components easily.

1. Strip about ¼ inch of wire from each wire on the two JST-PH plug connector cables.

2. Slip some heat-shrink tubing over the wires of one connector.

3. Connect the red wire of one cable to the red wire of the other cable by wrapping the bare wires together (Figure 6.6).

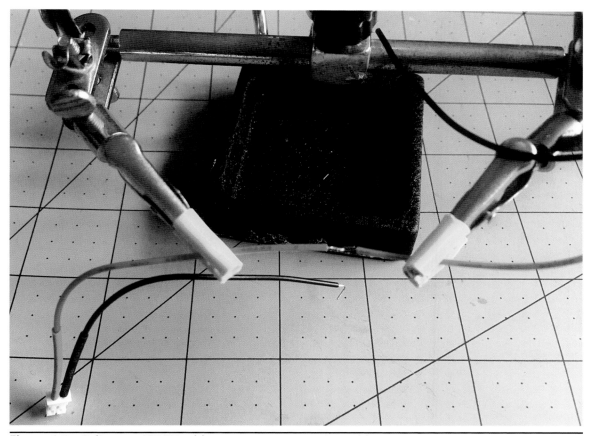

Figure 6.6   Splice two JST-PH cables to create a connection cable.

4. Solder the red wires, and repeat for the black wires.

5. Move the heat shrink into place, and apply heat (Figure 6.7).

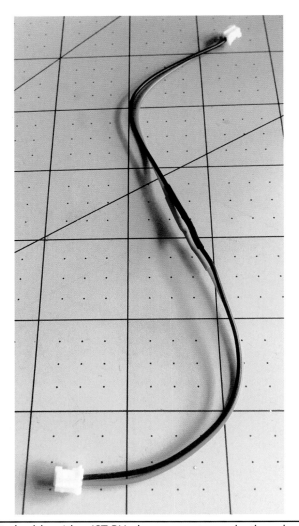

Figure 6.7 The completed cable with a JST-PH plug connector on both ends.

Step 5: Hook Everything Up

The central electrical component of this project is the solar charger. The solar panel plugs into it using the power jack labeled "DC IN." The battery plugs into it at the "BATT" socket. The Power Boost 500 is connected to the solar charger's "LOAD" socket via the connection cable (Figure 6.8).

When you plug in the solar panel, look for the red power-indicator light-emitting diode (LED) labeled "PWR" located near the Micro B USB port. If it turns on, the panel is providing power. If the battery is charging, the orange LED will turn on. When the battery is fully charged, the green LED will turn on. Both of these LEDs are located between the BATT and LOAD sockets.

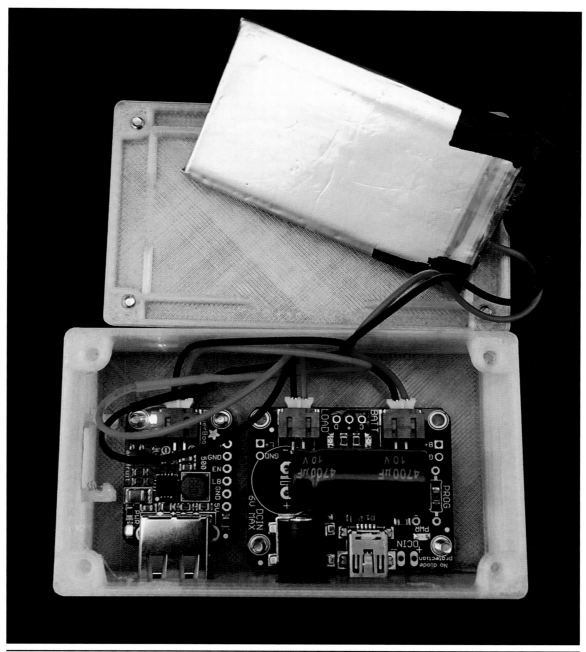

Figure 6.8 The battery connects to the BATT socket on the solar charger. The Power Boost 500 connects to the LOAD socket on the solar charger.

Step 6: Make the Backpack

Sew the backpack out of a sturdy canvas, denim, or outdoor fabric. Remember to backtack when stitching at the start and stop of all your sewing. For a no-sew option, purchase a simple backpack with a large top flap, and skip to the next step to attach your solar panel to the bag.

1. Cut out your back and front pieces, the two flap pieces, the pocket, and the two interfacing flap pieces.

2. Fuse the iron-on interfacing to the flap pieces. This will make the flaps stiffer and a better surface for mounting the solar panel (Figure 6.9).

Figure 6.9   Using an iron, fuse the interfacing to the fabric flaps.

3. Construct the interior pocket first. Make a hem on the top by turning $\frac{1}{4}$ inch on the upper edge of the pocket in to the wrong side and press, turn in again $\frac{1}{2}$ inch, and sew it down to finish the top edge.

4. Press the sides and bottom of the pocket $\frac{1}{2}$ inch in to the wrong side.

5. Pin the wrong side of the pocket to the wrong side of the back piece at the placement line approximately 4 inches from the top with the folded edges of the pocket facing in.

6. Stitch the pocket onto the back piece by sewing the sides and lower edge of the pocket close to the folded edge (Figure 6.10).

Figure 6.10   Sew the pocket onto the wrong side of the back piece.

7. Cut one piece of webbing 5 inches long, fold the end over $\frac{1}{4}$ inch, then $\frac{1}{4}$ inch again, and sew it with a straight stitch to make a nice, neat hem (Figure 6.11). We will put the hemmed end of the webbing through the prong side of the quick release clip in a later step.

Figure 6.11   Fold the edge twice, and sew it with a straight stitch to hem the webbing.

8. Pin the other end of the webbing piece you just sewed to the lower end of one flap section with wrong sides up and raw edges of webbing and flap even. Pin to hold the webbing in place (Figure 6.12).

Figure 6.12   Pin the edge of the webbing inside the flap.

9. Place the top flap pieces with their right sides together. The webbing should be sandwiched in between the flap pieces. With the right sides together, stitch the top flap sections together at the sides and lower edge. Trim the lower corners to make it easier to turn with thick fabric (Figure 6.13).

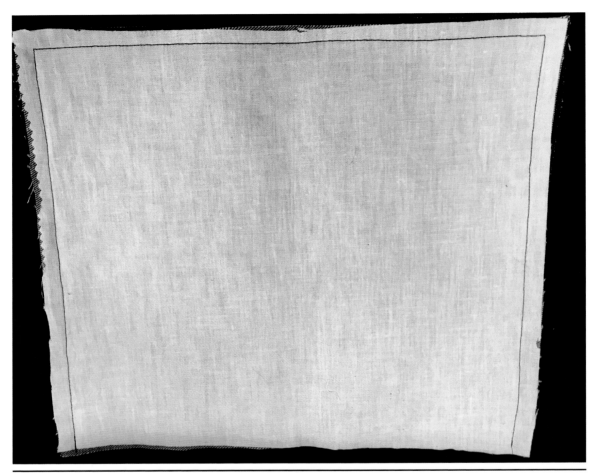

Figure 6.13   Sew the sides and lower edge of the flap.

10. Turn the top flap right side out, and press with an iron. The webbing should now be part of the flap and right-sides out. The top of the flap is still open. Turn the open edges in $\frac{1}{2}$ inch to face each other and pin them closed. Topstitch $\frac{1}{8}$ inch around flap (Figure 6.14).

Figure 6.14 Turn the open edges in ½ inch, and topstitch all the way around the flap.

11. Pin the flap to the back piece with the flap extending off the top of the back, along the line labeled "placement line for flap" on the pattern. Sew the flap to the back piece by stitching over the topstitching from the last step. After sewing the flap to the top back, insert the hemmed end of the webbing strap into the pronged end of the quick release clip.

12. Cut two pieces of webbing, each 12 inches long. On each piece, turn in ¼ inch on one end of the webbing piece, and press it with an iron. Turn in ¼ inch again and press. Stitch close to the inner pressed edge to make a neat hem.

13. Turn 1 inch on the lower end of the webbing piece to the opposite side and press. Do not sew it down. These will be the bottom backpack straps

14. Place the lower end of the webbing in the lower placement squares on the back piece with the 1-inch ends facing into the fabric. Stitch over the ends securely in a box and X shape, as shown (Figures 6.15 and 6.16).

Figure 6.15   Sew the straps securely to the backpack by stitching along an X and box shape.

Figure 6.16 Attach the back straps in the placement boxes on the back piece of the backpack.

15. Cut two pieces of webbing, 14 inches long. Fold one end of each over ¼ inch, then ¼ inch again, and then sew them with a straight stitch to make a nice, neat hem. Insert one end of each through the two D-rings. Fold the end down 1 inch, and stitch it closed close to D rings. Turn 1 inch on the lower end of the webbing piece to the opposite side and press. These are now the top straps

16. On combined back and flap pieces, place the lower end of the top strap webbing on the placement markings on the flap. Stitch it securely in a box and X shape as in step 14 (Figure 6.17).

Figure 6.17   Sew the top straps to the flap in the placement boxes.

17. Cut one piece of webbing 10 inches long. Fold one end over ¼ inch, then ¼ inch again, and then sew with a straight stitch to make a hem. The hemmed end of the webbing will be used for the other side of the quick release clip. Fold 1 inch down, and stitch closed close to the clip.

18. Pin the unhemmed end of the webbing piece to the center bottom right side of the front piece with the raw edges of webbing and the front even. Pin to hold the webbing in place (Figure 6.18).

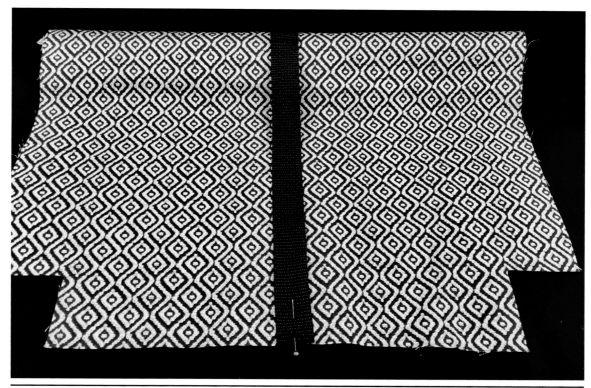

Figure 6.18   Attach the unhemmed side of the webbing to the center bottom of the front piece of the pack.

19. Pin the front to the back edges right sides together. The straps and the webbing piece should be sandwiched inside the bag. Sew along the bottom edge, and press the bottom seam open.

20. Fold the back and the front together again with the right sides together, and this time sew the front and back together at the sides, being careful to leave the small area between the arrows at the top open for inserting the drawstring later (Figure 6.19). Backtack on the seams where you are leaving space for the drawstring. Press the side seams open.

Figure 6.19 Sew the front and back pieces together along the side seams, leaving space for a drawstring.

21. Bring together the sides and bottom, matching seams. Sew across the edge to sew the corners closed on the bottom of the bag (Figures 6.20 and 6.21).

Figure 6.20 Bring the sides and bottom corners together, and sew across the edge to make the bottom corner of the bag.

Figure 6.21   Sew across the edge on the flattened corners.

22. Turn the whole bag right-side out. All the straps should now be on the outside.

23. Turn ¼ inch on the upper edge of backpack to the inside and press.

24. To form a casing for ribbon or cord, turn the edge inside again another inch and press; then stitch close to inner pressed edge (Figure 6.22).

Figure 6.22   Fold the top edge in twice, and stitch to make a drawstring casing.

25. Cut two pieces of cord, each 36 inches long, and insert one all the way around the casing using a safety pin to help pull it through. Both ends of one cord should stick out through the same hole. Knot the ends together.

26. Insert the other cord through the opposite casing hole, and pull all the way around. Knot the ends of the cord together (Figure 6.23).

27. On the front of the backpack, insert the remaining side of the quick release clip through the hemmed end of the webbing and make sure that it can clip together with the prong end on the flap. On the back of the backpack, insert the lower straps through the D-rings and adjust to fit.

Figure 6.23   Use two lengths of ribbon or cord to make a drawstring closure around the top of the bag.

Step 7: Attach the Solar Panel to the Backpack

1. Lay the top flap of your backpack flat. Place the solar panel evenly on the flap with the wires extending down toward the back of the bag. We will make five holes in the top flap, four for each of the solar panel attachment points and one for the wires to go into the bag.

2. Using chalk or pencil, make small marks on the flap to show the edges of the solar panel and also where the four attachment points are.

3. Unscrew the ends of the solar panel attachments. Use an awl to make small holes all the way through the flap that these points will fit into. Using a small sharp pair of scissors or the awl, make a hole in the flap fabric inside the rectangle of the solar panel so that the wires can go inside the bag (Figure 6.24). You can whipstitch the edges of this cut to make it neater.

Figure 6.24   Use an awl or small, sharp scissors to make holes in the flap to attach the solar panel and place the wires inside the bag.

4.   Insert the attachments points of the panel through the four holes in the flap, and screw the ends back on so that the solar panel is attached to the outside flap of the bag. Pull the connector wires through to the inside (Figure 6.25).

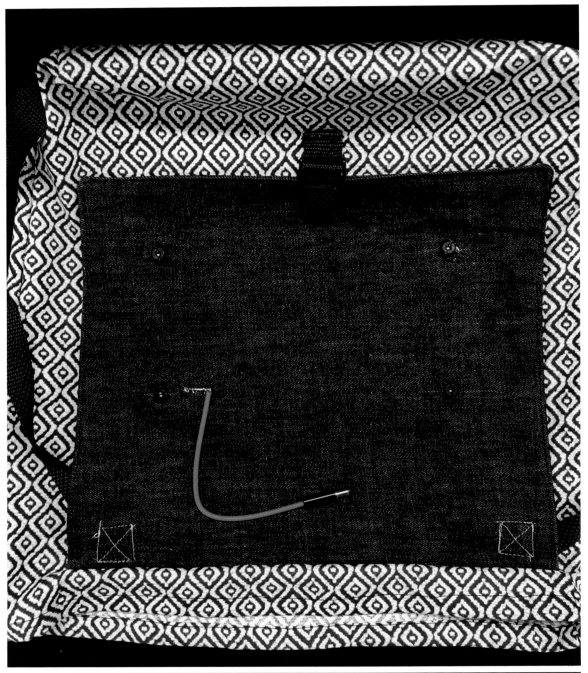

Figure 6.25   Insert the attachment posts through the holes, and put the screws on to hold the panel. Then pull the connector wires through the slit.

Step 8: Use It Outside

Place the electronics and your phone inside the pocket on the bag. Take the bag out to direct sunlight to charge your phone! Remember, the solar panel works best pointing toward the sun, preferably perpendicular to the sun. Caution: Solar panels can become hot when charging in direct sunlight. Moving the panel to shade or indoors should quickly lower the temperature if the panel ever becomes very hot.

CHAPTER 7

Starlight Skirt

MAKE A STYLISH SKIRT AND PROGRAM IT USING THE ARDUINO IDE TO SHINE IN ALL the colors of light.

What You Will Learn

In this chapter, you will learn how to

- Work with loose fiber-optic strands
- 3D print fiber-optic holders and light-emitting diode (LED) diffusers
- Arduino programming

Files You Will Need

- Star-shaped diffuser models for 3D printing: starDiffuser.stl
- Starlight skirt Arduino sketch: starlightSkirt.io

Tools You Will Need

- Soldering tools: soldering iron and solder
- 3D printer
- Hot-glue gun
- Sewing machine and thread
- Arduino programming environment

Materials You Will Need

- **3D printer filament:** polylactic acid (PLA), acrylonitrile butadiene styrene (ABS), glow-in-the-dark nylon

- WS2812b addressable LEDs (One meter of prewired LEDs is recommended.)

- JST-SM connectors (only if your addressable LEDs don't have the correct connector)

- 50 to 100 meters of 0.5-, 0.75-, or 1.0-millimeter end-glow fiber-optic strands

- Craft scissors

- Hot-glue gun and glue sticks or other glue suitable for plastics

- Electrical tape (available in different colors)

- Heat shrink

- Fiber-optic diffusers

- StitchKit MakeFashion controller board (www.stitchkit.io), or other small Arduino-compatible board

- Battery, either a USB battery pack or a LiPo battery

- 1.5 to 2 yards of fashion fabric (wovens recommended)

- 1 yard of lining material for lining and pocket: cotton, cotton/poly, other woven fabrics

- 2 hooks and eyes ⅝ inch wide, usually called *skirt* or *pants hooks*

- Matching thread

- **Optional:** Lightweight iron-on interfacing for the waistband of the skirt

- **Low-sew option:** Purchased knee- or above-knee-length skirt with a flat waistband

The Starlight skirt has easy-to-connect electronics and a bit of programming. This time we'll be using the Arduino programming environment, also known as the Arduino IDE. The electronics are the StitchKit Fashion Technology Kit, which was designed by fashion tech makers to work with wearable projects. Inside the StitchKit you'll find the MakeFashion board and most of the components you need to make this project (Figure 7.1). Other small Arduino-compatible boards such as those made by Arduino, Sparkfun, or Adafruit can be substituted for the StitchKit MakeFashion board.

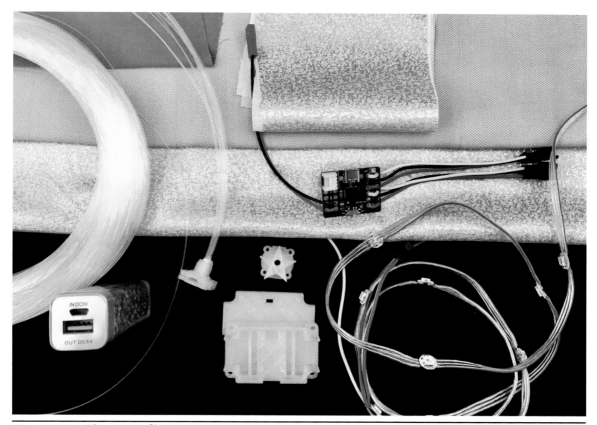

Figure 7.1 Fiber-optic filament, wired LEDs, and MakeFashion board are used for the Starlight skirt.

Step 1: Sew the Skirt

Remember to sew a backtack at the start and stop of your seams.

1. You will need two measurements for this skirt. First, measure your waist where you want the waistband of the skirt to sit. Then measure how long you want the finished skirt, for someone who is 5 feet, 6 inches tall, somewhere between 18 and 22 inches will make an above-the-knee or knee-length skirt, respectively.

2. Cut the lining fabric. Cut a rectangle that is as long as 1.5 times your waist measurement with a width of how long you want your finished skirt.

Lining = length (1.5 × waist measurement) × width (desired length of skirt)

3. Cut the skirt fashion fabric. Cut a rectangle that is as long as 2.5 times your waist measurement with a width of the how long you want your skirt + 4 inches. If you want a really full skirt, you can cut a rectangle that is 3 times your waist measurement. Depending on how wide your fabric is, you may need to sew two rectangles together to get this length.

$$\text{Skirt fabric} = \text{length } (2.5 \times \text{waist measurement})$$
$$\times \text{ width (desired length of skirt} + 4 \text{ inches)}$$

Or for a very full skirt,

$$\text{Skirt length} = \text{length } (3 \times \text{waist measurement})$$
$$\times \text{ width (desired length of skirt} + 4 \text{ inches)}$$

4. Cut the waistband. The fabric is the same as your skirt fashion fabric. Cut a waistband that is a rectangle as long as your waist measurement + 4 inches for the seam allowances and 4 inches wide. The skirt will overlap at the back. If you want to add interfacing to make your waistband a little stiffer, cut the same measurements out of a lightweight iron-on interfacing. Iron the interfacing to the waistband before you start sewing.

$$\text{Waistband} = \text{length (waist measurement} + 4 \text{ inches)} \times \text{ width (4 inches)}$$

5. Cut a pocket. This pocket will go inside the skirt next to the waist opening and will hold the microcontroller board and battery. Cut the pocket out of lining fabric.

$$\text{Pocket} = \text{length 10 inches} \times \text{width 6 inches}$$

6. Make the pocket. To hem the pocket, fold the 6-inch side in $\frac{1}{4}$ inch, press, and then fold $\frac{1}{4}$ inch again. Sew along the fold to hem the pocket. Fold the hemmed end of the pocket up 4 inches. Sew along both sides of the pocket with a $\frac{1}{2}$-inch seam allowance. Turn the pocket inside out so that the raw edges are on the inside (Figure 7.2).

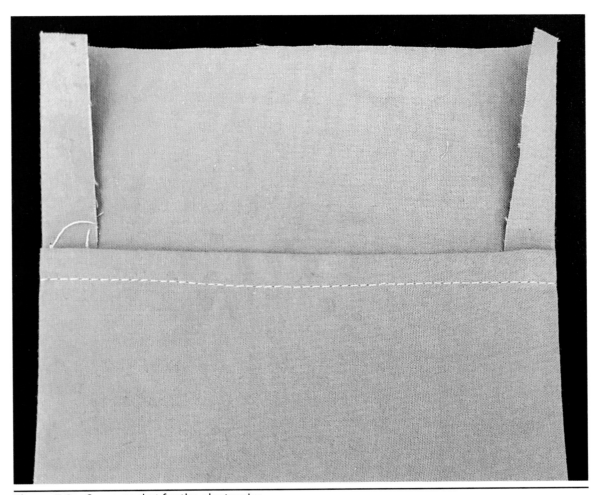

Figure 7.2   Sew a pocket for the electronics.

7. Sew your side seams, remembering to backtack at the start and stop of each seam. Fold the lining piece so that the short ends are together. Sew with a $\frac{1}{2}$-inch seam allowance, leaving the top 5 inches open. This opening is how you get into the skirt. Turn the seam allowance under on each side of the opening, and sewn it down to make a neat hem. Fold your fashion fabric piece so that the short ends are together, and sew with a $\frac{1}{2}$-inch seam allowance, leaving the top 5 inches open. Finish the opening as you did with the lining piece (Figure 7.3).

Figure 7.3 Sew down the sides of the back opening to make a neat hem.

8.   You will need to gather the material in order to sew the skirt together.

   a.   Start the gathering with a long thread tail for your needle and bobbin thread. Do not backtack when you start and stop sewing gathers.

   b.   Using the longest straight stitch on your machine, sew about ¼ inch away from the raw edge of the fabric all the way. Leave a long thread tail when you are done.

   c.   Sew another line of long, straight stitches about ⅜ inch away from the edge. Having two lines of stitches prevents the thread from breaking as you gather the fabric.

   d.   Wind the bobbin threads around your fingers to create gathers in the fabric. Gently keep pulling and positioning the gathers until you get the length of fabric you want. Once you get the right length, you can tie a knot in the fabric tails to keep everything in place (Figure 7.4).

Figure 7.4   Gather the fabric along the stitching.

9. Gather the top side of the lining with the opening into the length of the waistband (waist measurement + 4 inches).

10. Gather the top side of the skirt fabric with the opening into the length of the waistband. Gather the bottom side of the skirt fabric into the length of the lining (1.5 × waist measurement).

11. Sew the bottoms together. With the right sides facing each other, pin the bottom edge of the lining to the bottom edge of the gathered skirt fabric. Sew with a ½-inch seam allowance (Figure 7.5).

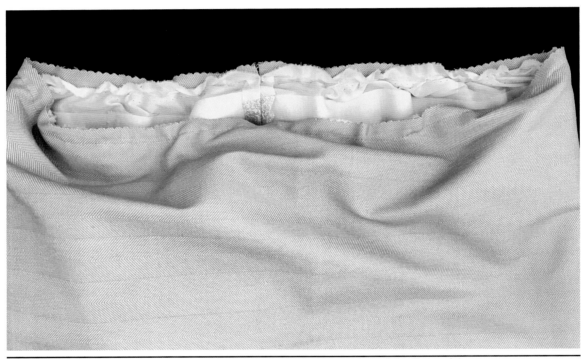

Figure 7.5   Sew the bottom of the skirt fabric to the bottom edge of the lining fabric.

12. Press this seam, and then turn the fabric out so that the seam is on the inside of the skirt.

13. Pin the top of the skirt fabric to the top of the lining fabric, matching up the openings at the back. Sew these pieces together with a straight stitch ¼ inch from the raw edges at the top (Figure 7.6).

Figure 7.6   Sew the skirt and lining together at the top.

14. Sew the pocket to the inside of the skirt. Pin the raw edge of the pocket to the inside of the skirt and lining pieces 1 inch from the openings. Sew to the skirt piece with a straight stitch ¼ inch from the top (Figure 7.7).

Figure 7.7   Sew the electronics pocket to the inside of the skirt.

15. Fold the waistband in half lengthwise with the right sides together. Sew each of the short ends with a $\frac{1}{2}$-inch seam allowance. Turn the right side out, and press (Figure 7.8).

Figure 7.8   Sew the ends of the waistband and turn the right sides out.

16. Attach the waistband to the skirt. Match up one long edge of the waistband with the raw edges of the skirt. Sew the waistband on with a ½-inch seam allowance (Figure 7.9).

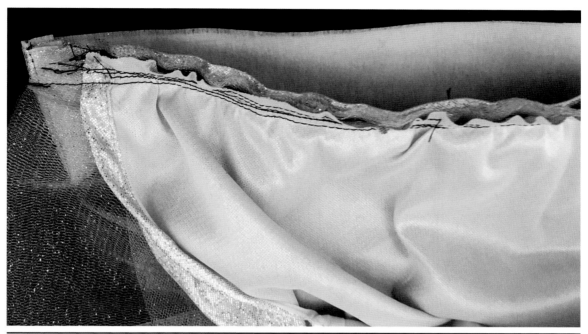

Figure 7.9 Sew one side of the waistband to the skirt.

17. Press the waistband open so that the raw edges of the skirt go inside the folded waistband. You will now have one long edge of the waistband that isn't sewn yet. Fold the long edge of that side in ½ inch, and pin it so that it covers the stitching that attached the skirt (Figure 7.10).

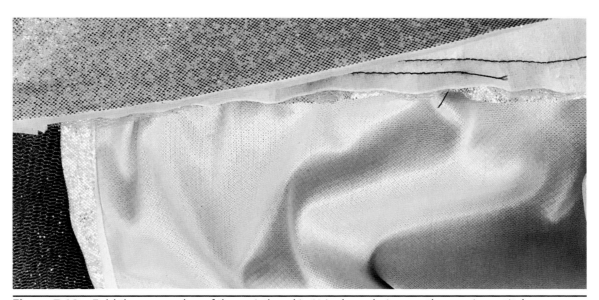

Figure 7.10 Fold the open edge of the waistband in ½ inch, and pin over the previous stitches.

18. Sew the waistband closed by hand or with a machine, enclosing the raw edges of the skirt completely in the waistband.

19. Use the hooks and eyes to close the waistband. Sew the eyes on $\frac{1}{2}$ inch from the edge of the waistband. Sew the hooks on $\frac{1}{2}$ inch from the edge of the overlap (Figure 7.11).

Figure 7.11   Use large hooks and eyes to close the skirt waistband.

Step 2: 3D Print the Fiber-Optic Holders and LED Diffusers

1. Using the waistband measurement for your skirt, measure the wired LEDs, leaving enough wire to connect to the board, and place the board in the skirt pocket.

2. Count the number of LEDs on your strand. This is the number of star diffusers you will need for the skirt. You will need both the snap-on base and the fiber-optic holder tops. Both are included in the downloadable model (Figure 7.12).

3. Download the model of the starDiffuser.stl file, and open it in Cura for slicing. The model includes the snap-on base.

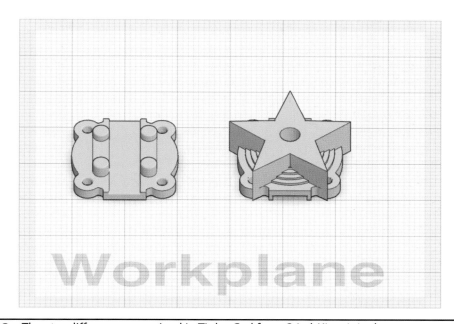

Figure 7.12   The star diffuser was remixed in TinkerCad from StitchKit originals.

4. Load the filament, and slice the model with the following settings: 10 percent infill; Generate Support: yes; Build Plate Adhesion Type: Brim. Start your 3D printer according to the operating instructions. Print!

5. On a filament printer the diffuser tops will print with thinner plastic support material inside. Remove this material with pliers by twisting and pulling (Figure 7.13).

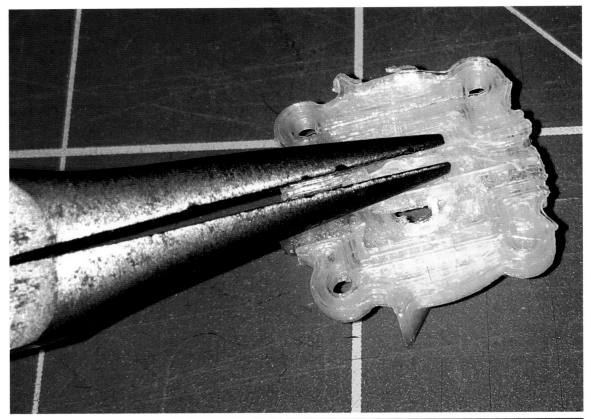

Figure 7.13   Use pliers to remove the printed plastic supports inside the diffuser tops by twisting and pulling.

Step 3: Put the Fiber Optics Together

1. If your optical fiber came in a long roll, you will need to cut the fiber into strands using scissors. Strands should be between 1 to 2 feet long, and you will probably need seven to ten strands per diffuser.

2. Bundle the fibers together, holding the ends flush. Insert the bundle through the star diffuser to determine how many fibers can fit in each bundle. Remove the bundle.

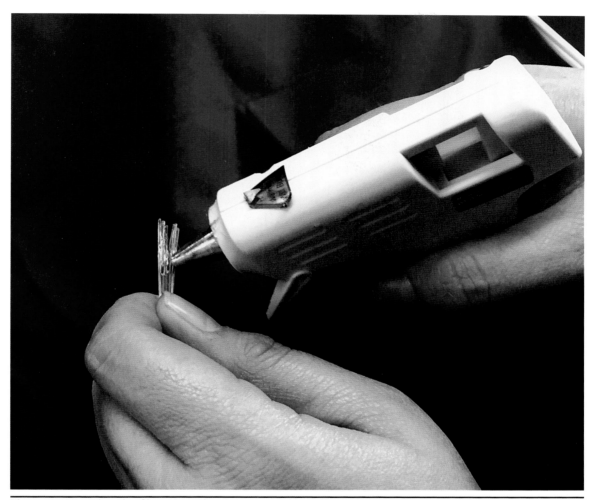

Figure 7.14 Glue the fibers together into a round bundle.

3. Use a small amount of glue between $\frac{1}{8}$ and $\frac{1}{2}$ inch from the end of the bundle to stick the fibers together. Allow the glue to dry. You can also wrap the bundle in electrical tape, although it will not be as secure as with the glue (Figure 7.14).

4. Once your bundle is dry, recut the end so that all the strands are flush. The flatter the ends are, the better they will make contact with the LED and transmit light (Figure 7.15).

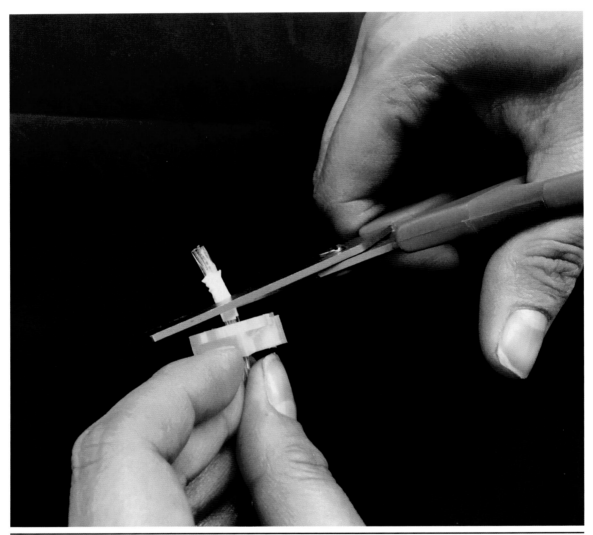

Figure 7.15 Cutting the optical fiber into 1- to 2-foot lengths. Try to keep the ends flat.

5. Push the bundle back into the diffuser, and line it up with the holder, leaving enough room for the LED and diffuser base to snap together. Test it to be sure that everything fits; then unsnap it. To secure the fiber optics in the diffuser, apply hot glue or other glue along the outside edge (Figure 7.16).

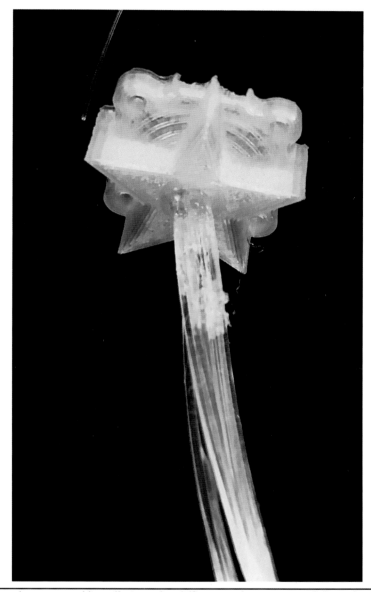

Figure 7.16 Insert the wrapped bundle into the diffuser, and secure it with hot glue.

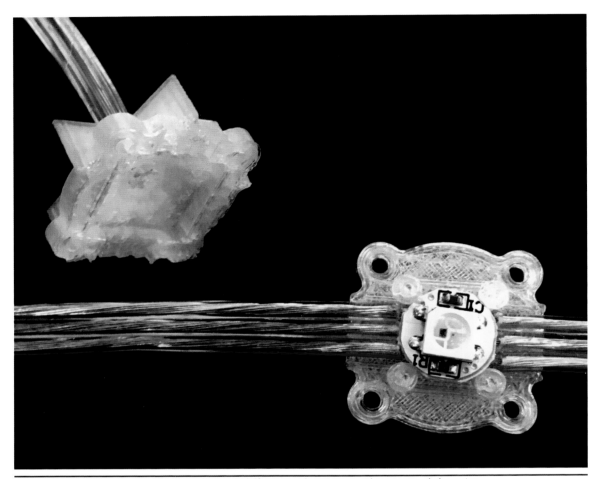

Figure 7.17   Wired LEDs fit into the star diffuser and snap together around the wire.

6.   Attach the diffusers to the LED strip (Figure 7.17).

7.   Sew the LED diffusers onto the waistband using the holes on each corner of the diffuser. You may want to sew the bases down first and then snap the two parts together and sew the whole diffuser to the waistband with a whipstitch through the holes in the corners (Figure 7.18).

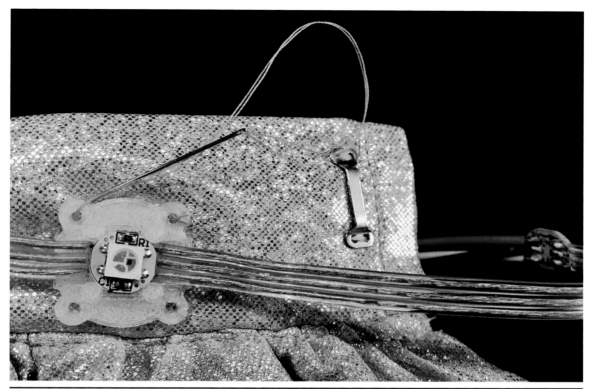

Figure 7.18   Sew the diffuser via the corner holes onto the waistband with a whipstitch.

Step 4: Connect the Electronics

1. If needed, add connectors to the addressable LED strips. The StitchKit controller board comes with a JST-SM plug connector that can be used to connect to WS2812b addressable LED strands (Figure 7.19).

2. If your stand of addressable LEDs does not have a socket JST-SM connector, you can remove the old connecter and solder on a new one, or you can create a short adapter cable. If you are relatively new to soldering, we recommend creating a short adapter cable.

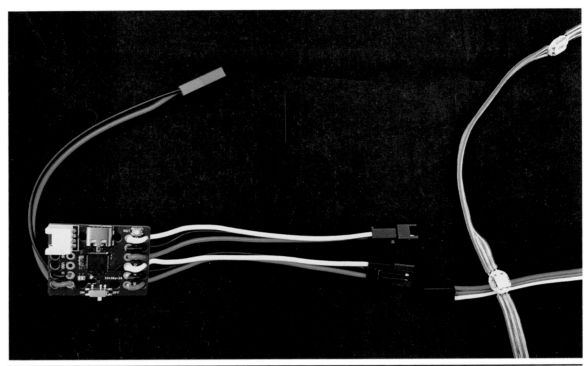

Figure 7.19   LED strips will need a JST-SM socket connector in order to plug into the StitchKit controller board.

3. Create a short adapter cable (if needed). If your LED strip came with the same JST-SM plug connector as the StitchKit's MakeFashion board, create a short adapter cable consisting of two JST-SM socket connectors. Solder the red wire to the red wire and the black wire to the black wire. Use heat shrink or electrical tape to insulate and protect your solder connections (Figure 7.20).

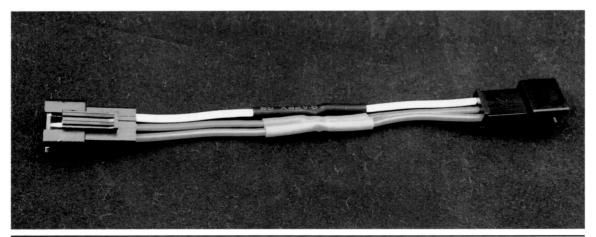

Figure 7.20   Sometimes addressable LED strips come with a plug connector. You can remove the old connecter and solder on a new one, or you can create a short adapter cable.

Step 5: Program the StitchKit's MakeFashion Board

The StitchKit's MakeFashion board can be programmed using the Arduino IDE (Integrated Development Environment). The Arduino IDE lets you create programs called sketches and upload them to StitchKit's MakeFashion board. Unlike the MakeCode programming environment we used to program the hip pack (Chapter 5), the Arduino IDE is an application (app) that you have to install on your computer.

1. Download the Arduino IDE through the Arduino website (www.arduino.cc/en/Main/Software). There are two choices, the Arduino Web Editor, which lets you work online but requires a reliable Internet connection, and the Arduino Desktop IDE. We'll be using the Desktop IDE, which lets you work offline on your computer.

2. Download the NeoPixel Library by accessing the Library Manager by choosing "Manage Libraries" from the "Include Library" menu item under the "Sketch" menu (Figure 7.21). Search for "Adafruit NeoPixel," and install it (Figure 7.22).

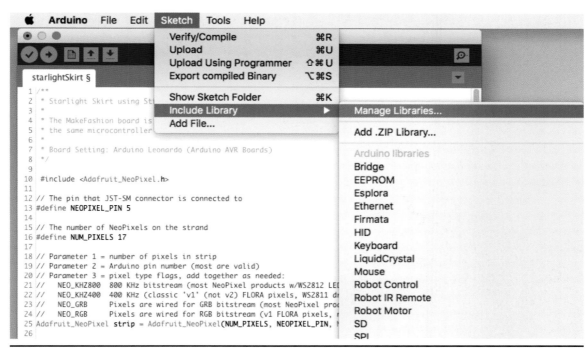

Figure 7.21 Access the Library Manager from the "Include Library" menu item under the "Sketch" menu.

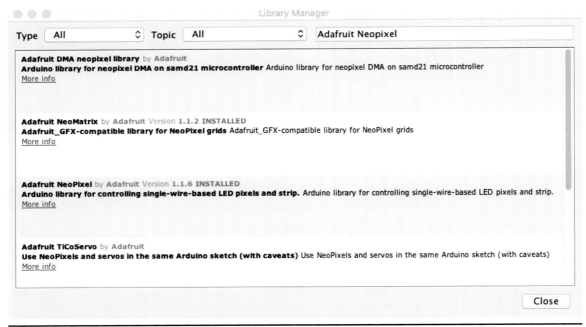

Figure 7.22   Search for "Adafruit Neopixel" and install it from the Library Manager.

3. Download the starlightSkirt folder contain the starlightSkirt.io sketch and open the sketch in the Arduino IDE.

```
#include <Adafruit_NeoPixel.h>
#define NEOPIXEL_PIN 5
#define NUM_PIXELS 17
Adafruit_NeoPixel strip = Adafruit_NeoPixel(NUM_PIXELS, NEOPIXEL_PIN,
  NEO_GRB + NEO_KHZ800);
void setup() {
  strip.begin();
  strip.setBrightness(100); //adjust brightness here
  strip.show(); // Initialize all pixels to 'off'
}
void loop() {
    rainbowCycle(20);
}
void rainbowCycle(uint8_t wait) {
  uint16_t i, j;
  for(j=0; j<256*5; j++) { // 5 cycles of all colors on wheel
    for(i=0; i< strip.numPixels(); i++) {
      strip.setPixelColor(i, Wheel(((i * 256 / strip.numPixels()) + j)
        & 255));
    }
    strip.show();
    delay(wait);
  }
```

```
}
uint32_t Wheel(byte WheelPos) {
  if(WheelPos < 85) {
   return strip.Color(WheelPos * 3, 255 - WheelPos * 3, 0);
  } else if(WheelPos < 170) {
   WheelPos -= 85;
   return strip.Color(255 - WheelPos * 3, 0, WheelPos * 3);
  } else {
   WheelPos -= 170;
   return strip.Color(0, WheelPos * 3, 255 - WheelPos * 3);
  }
}
```

4. Update the program to reflect the number of LEDs in your project. Count the number of LEDs you are using for your skirt. To get the controller to light that number, change the code in the third line "#define NUM_PIXELS 17" from 17 to however many LEDs are on your skirt.

5. Select "Arduino Leonardo" from the "Board" menu item under the Tools menu. The StitchKit controller board is based on the ATmega32U4 microcontroller, the same microcontroller used in the Arduino Leonardo board (Figure 7.23).

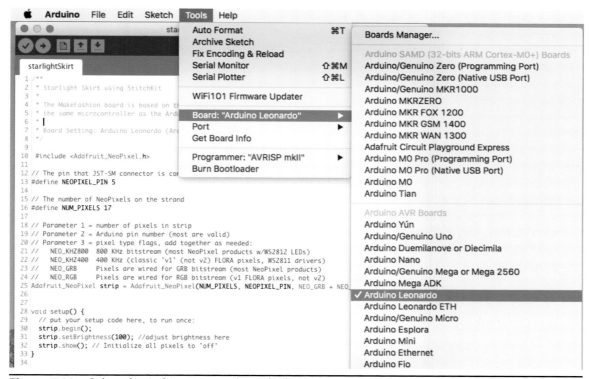

Figure 7.23   Select the Arduino Leonardo as the board type.

6. Using the USB cable that came in the StichKit, plug the StitchKit's MakeFashion board into your computer.

7. Select the board's serial port from the "Port" menu item under the Tools menu. On the Mac, the serial port is probably something like /dev/tty.usbmodem241. On Windows, it's a COM port.

8. When you're ready to upload the sketch, select "Upload" from the "Sketch" menu. Your skirt will down cycle through the colors of the rainbow (Figure 7.24).

Figure 7.24   Starlight skirt showing the fiber-optic diffusers. (*Courtesy of Jason Martineau*)

Step 6: Going Further

It's hard to beat a cycling rainbow pattern, but with the StitchKit MakeFashion board you can also add sensors so that your skirt can react to sound, movement, and more. In fact, the StitchKit MakeFashion board works with an assortment of Seeedstudio's Grove sensors and inputs such as accelerometers, loudness sensor, light sensors, and buttons. The important thing is to have fun (Figure 7.25).

Figure 7.25 The Starlight Skirt is a fun project to wear.

CHAPTER 8

Programmable Sewn Circuit Cuff

FELT, LEATHER, AND SEWN CIRCUITS GIVE THIS MODERN TECH PROJECT AN ORGANIC FEEL. You can choose to make a plain cuff with the sewn circuit or use one of the laser-cutter files to use the laser to engrave a design on your cuff.

What You Will Learn

In this chapter, you will learn how to

- Sew a project with conductive thread
- Use a capacitive touch sensor to change lighting programs
- Program the Adafruit Gemma M0 in CircuitPython

Files You Will Need

- Python code: code.py
- Cuff pattern: heart_cuff.pdf
- Cuff pattern for laser cutter: heart_cuff.svg
- Optional battery holder 3D model: lipo_sleeve.stl

Tools You Will Need

- Sewing tools and supplies
- Computer
- **Optional:** Laser cutter

- **Optional:** Leather-working tools, including a leather needle and leather punch
- **Optional:** 3D printer for printing a protective battery case

Materials You Will Need

- Gemma M0 (www.adafruit.com/product/3501)
- Four or more sewable addressable light-emitting diodes (LEDs; WS2812), Adafruit Flora NeoPixels v2 (www.adafruit.com/product/1260)
- Stainless steel conductive thread:
    - 30 feet of stainless steel thread (www.sparkfun.com/products/10867)
    - Or 60 feet of three-ply stainless steel thread (www.adafruit.com/product/641)
- Small rechargeable lithium polymer (LiPo) battery (www.adafruit.com/product/1317)
- Small USB battery charger (www.adafruit.com/product/1304)
- USB data cable for programming the Gemma M0 (Some USB cables are only for power; this requires one that is data and power.)
- Fray Check sealant, fabric glue, or clear nail polish to seal conductive thread knots
- Cuff material:
    - 5- × 10-inch (or larger) piece of wool or wool-blend felt (0.5 to 3 millimeters thick; available at JOANN stores or online retailers)
    - Or 1.5 to 3 ounces of thin vegetable-tanned leather (0.3 to 1.2 millimeters thick; available at Tandy Leather, in store or online [www.tandyleather.com/en/])
    - Optional: Eco-Flo leather dye in preferred color (if you want to dye your leather)
- Sewing needle with large eye or leather needle (if you are making your cuff out of leather)
- Two plastic or metal snaps (sew-on size 4 recommended) or other fasteners (buttons, studs, etc. as preferred) (Do not use brass or nickel-plated fasteners because they could be conductive and can cause short circuits with your sewn circuit.)
- **Optional:** A larger capacitive touch sensor such as a small ($\frac{1}{2}$ inch or less) silver or silver-plated charm to add to the capacitive touch sensor
- **Optional:** 3D printed battery case

This project uses an Adafruit Gemma M0 microcontroller and Flora NeoPixel addressable LEDs (Figure 8.1). The Gemma M0 is the third iteration of Adafruit's tiny sewable Gemma microcontroller board. It's only about 1 inch in diameter, yet it includes a 32-bit SAMD21 microcontroller, a USB connector, a battery connector, an on/off switch, and three pins of the microcontroller broken out to large sewable pads.

NeoPixels is Adafruit's brand name for addressable LEDs controlled by the WS2811, WS2812, and WS2812B controller chips. These chips are paired with each LED and allow individual addressing of daisy-chained LEDs. NeoPixels can be controlled from a single pin on the microcontroller. They are directional, meaning that they need to be wired in a particular orientation. The Flora NeoPixels are about ½ inch in diameter and also have large holes on conductive pads that allow them to be easily sewn with conductive thread. Tips and tricks for sewing with conductive thread are covered in Chapter 2.

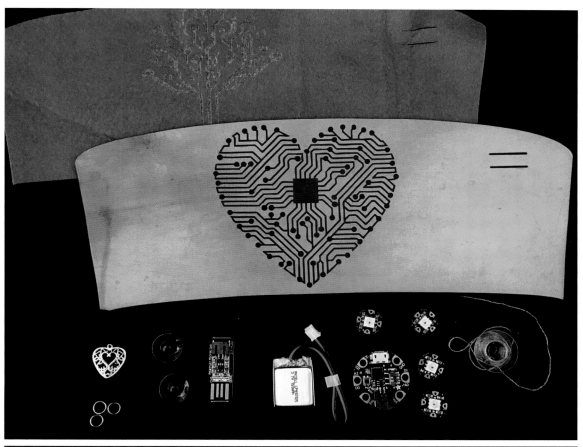

Figure 8.1 Electronic components for the color-change cuff: sewable addressable LEDs, the Gemma M0 microcontroller board, the LiPo battery, and the small USB charger.

Working with Leather

A leather cuff is a great fashion-forward accessory or steampunk costume piece. There are a few tools and tips specific to working with leather. The laser cutter can etch beautiful designs into vegetable-tanned leather; other types of leathers may not etch as well and may contain dangerous materials that can release toxic gases when burned with a laser.

Dye the leather with water-based dye such as Eco-Flo before engraving the design on the laser cutter. Dye can be purchased at your local leather supply or craft store. The major difference between making a leather cuff and a felt cuff is that you will need to prepunch holes to sew the conductive thread through the leather. When you are constructing your cuff and need to sew regular or conductive thread, you will need to use an awl or leather punch to make the hole before stitching with the needle and any kind of thread (Figure 8.2). Use a leather needle to sew both regular sewing thread and conductive thread.

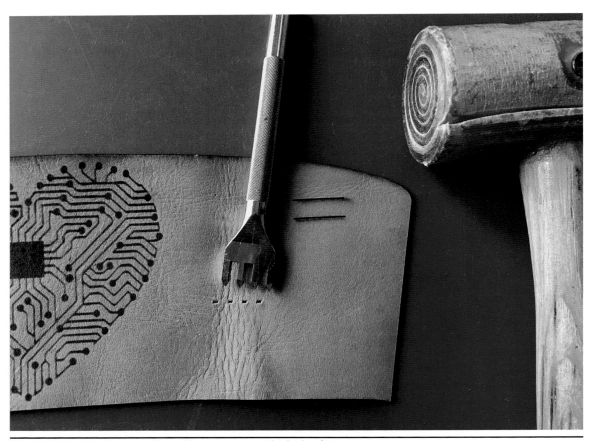

Figure 8.2   Leather punch and hammer to make holes for sewing.

Step 1: Make the Cuff

1. Download the cuff template as a .pdf file or download the laser engrave files.

2. **No-laser option:** Using sharp scissors, a rotary cutter, or a hobby knife, cut your cuff out of felt or leather.

3. **Laser option:** Before cutting the laser file, you will need to tell your laser which parts of the image should be engraved and which lines should be cut. The engraved images in the download file are raster graphics and are in black; the cutting lines are vector graphics and are in red (Figure 8.3). Consult your laser operating instructions for the colors and line weights that are associated with raster and vector graphics. Set up your file to cut on the laser, and consult the laser operating recommendations for the correct power and cutting speed for fabric or leather depending on your material choice. Engrave and cut the cuff.

Figure 8.3   The laser engraving file uses color differentiation between laser engraving and laser cutting.

Step 2: Charge the Battery

When you first buy LiPo batteries, they will need to be charged. In this project, we are using the Adafruit Micro Lipo Charger, a small, affordable battery charger. On one side the LiPo rechargeable battery plugs into the JST-PH socket. On the other side are four long gold-plated contacts that plug into any USB port, like those on a computer or phone charger. While charging, the red LED is lighted. When the battery is fully charged and ready for use, the green LED turns on.

Be careful when using LiPo batteries. Never bend, puncture, or crush them. Only use a battery charger designed for LiPo batteries. Never use a charger meant for NiMH, NiCad, or lead-acid batteries!

Step 3: Lay Out and Sew the Electronics

1. Lay out the electronics. With the pattern pieces cut out, decide how you are going to lay out the electrical components. The Flora NeoPixels have two arrows on top of the board that show the in and out directions. As you orient your neopixels, ensure the arrows are all pointing in the same direction when they are wired together. The Flora NeoPixels require three pins from the Gemma M0: (1) voltage out, (2) ground, and (3) single data pin. These are all conveniently located by the power slide switch on the right side of the Gemma M0. Don't forget to leave room for the battery on the other side of the Gemma (Figure 8.4).

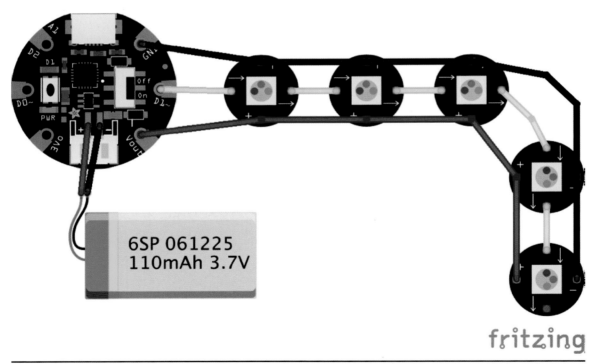

Figure 8.4   The NeoPixels are daisy-chained on the power switch side of the Gemma M0 board.

2. Once you are happy with where the NeoPixels are located on your cuff, you can start sewing on your board and components. Sew the components onto the cuff with regular sewing thread before making the sewn circuit because sewing thread is stronger than conductive thread and less prone to fraying. Whipstitch once or twice through the larger copper connection holes on the Gemma. Don't cover the copper entirely because you want to leave plenty of room for your conductive thread to make a good connection. Take one whipstitch with regular sewing thread through each of the holes on the NeoPixels.

3. Using a running stitch, sew the voltage connections with conductive thread. The voltage line, marked with a plus sign (+), will be sewn continuously from the Gemma through each Flora NeoPixel. This means that you can use one long thread to connect each NeoPixel and do not have to cut separate threads for each voltage connection. Starting at the copper pin marked "Vout," sew through each plus sign (+) on each NeoPixel, remembering to whipstitch numerous times and tightly around the copper pads on each component to get a good connection and to make tight knots when you start and stop sewing.

4. Using a running stitch, sew the ground connections with conductive thread. The ground connection, marked with a minus sign (−), will be sewn as a separate continuous line from the Gemma M0 through each Flora NeoPixel. Starting at the copper connector marked "GND," sew continuously through each minus sign (−) on the NeoPixels.

5. Using a running stitch, sew the data connections with conductive thread. Separate connections are sewn from "data out" to "data in" on each NeoPixel. Starting at the D1 pin, sew to the in arrow on the first NeoPixel. Whipstitch through the hole, and tie a tight knot. Then cut your conductive thread to sew the next connection. Starting at the out arrow, knot your thread, whipstitch around the component, and stitch to the in arrow of the next NeoPixel. You may want to take a couple backstitches in the beginning to be sure that the start is tightly connected. Repeat for all NeoPixels (Figure 8.5).

6. After sewing the conductive thread, trim the thread tails on the knots to ensure you won't get a short circuit. To help keep your knots in place, put a small dot of Fray Check sealant, fabric glue, or clear nail polish on all the conductive thread knots.

7. Finish your cuff by sewing on snaps, buttons, or other fasteners to hold it securely around your wrist.

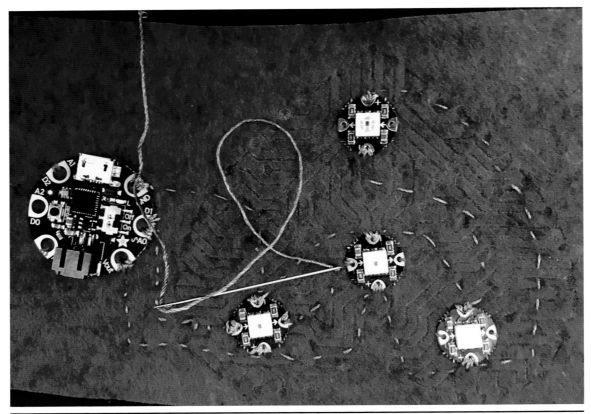

Figure 8.5   The components connected with conductive thread on felt. The components are whipstitched in place with red thread.

Step 4: Program the Gemma

The Gemma M0 can be programmed using either CircuitPython or the Arduino IDE. Because the Gemma M0 ships with the CircuitPython firmware installed, we'll be using the CircuitPython. It's an extension of MicroPython with support for Adafruit products. The CircuitPython firmware mounts the Gemma M0 as a small USB drive on your computer, allowing you to create, edit, and store your program directly on the device. CircuitPython will detect when a program file has been saved and automatically run the program. CircuitPython looks for a file named code.txt, code.py, main.txt, or main.py in this order and will run the first one it finds.

1.   Download the Python program. It will be named code.py and looks like this:

```
import board
import digitalio
import neopixel
```

```python
import random
import time
import touchio

led = digitalio.DigitalInOut(board.D13)
led.direction = digitalio.Direction.OUTPUT
pixels = neopixel.NeoPixel(board.D1, 4, brightness=0.1)
touch1 = touchio.TouchIn(board.A1)
myColors = [(232, 100, 255), (200, 200, 20), (30, 200, 200)]

def wheel(pos):
  if (pos < 0) or (pos > 255):
    return [0, 0, 0]
  if (pos < 85):
    return [int(pos * 3), int(255 - (pos*3)), 0]
  elif (pos < 170):
    pos -= 85
    return [int(255 - pos*3), 0, int(pos*3)]
  else:
    pos -= 170
    return [0, int(pos*3), int(255 - pos*3)]

def flashRandom(wait):
  aColor = random.choice(myColors)
  red = aColor[0]
  green = aColor[1]
  blue = aColor[2]
  randPixel = random.randrange(0, len(pixels))
  for j in range(5):
    r = red * (j+1)/5
    g = green * (j+1)/5
    b = blue * (j+1)/5
    pixels[randPixel] = (int(r), int(g), int(b))
    pixels.write()
    time.sleep(wait)
  for j in range(5):
    k = 4-j
    r = red * (k)/5
    g = green * (k)/5
    b = blue * (k)/5
    pixels[randPixel] = (int(r), int(g), int(b))
    pixels.write()
    time.sleep(wait)

def rainbow_cycle(iColor, wait):
  for i in range(pixels.n):
    idx = int((i * 256 / len(pixels)) + iColor)
    pixels[i] = wheel(idx & 255)
```

```
      pixels.write()
      time.sleep(wait)

shouldTwinkle = True
iColor = 0
while True:
    if touch1.value:
        led.value = not led.value
        shouldTwinkle = not shouldTwinkle
        time.sleep(0.25)
        pixels.fill((0, 0, 0))
    if (shouldTwinkle is True):
        flashRandom(0.02)
    else:
        iColor = (iColor+1) % 256
        rainbow_cycle(iColor, 0.001)
```

2. Plug the Gemma M0 in to your computer using a USB data cable. Turn the Gemma M0 on using the power switch on the board. The built-in status light might be green, indicating that a program is running, or it might be swirling through the colors of the rainbow in a demo program. Either way, the Gemma M0 should mount as a small 64K flash drive.

3. Copy the code.py file to the Gemma M0. After the file has finished copying, the program should run, randomly flashing the NeoPixels.

4. Make sure to eject the Gemma M0 before unplugging it from a computer.

Step 5: Use the Capacitive Touch Sensor

Capacitive touch sensors can measure the electrical properties of the human body. This is similar to how the touchscreen on your phone or computer can tell where your finger is touching. The Gemma M0 can measure capacitive touch on the copper pin pads, and we will use these to change between the LED patterns that are programmed on the cuff.

1. Be sure that you have loaded the code.py program onto your Gemma M0.

2. Turn the Gemma M0 on using the on/off switch. Briefly touch the A1 pad to change the pattern.

3. **Optional:** Because the A1 pad is so small, you can add a conductive silver jewelry charm to make the sensor a little bigger. You don't want a big charm because that might accidentally cause a short circuit. Whipstitch the charm to pad A1 with conductive thread, just going through the charm and the pad. Do not sew all the way through the

cuff because then the thread will touch your wrist and constantly activate the sensor. Keep a tight connection between the charm and the A1 pad. To keep the charm from banging around, whipstitch the other end of the charm down with regular sewing thread. If you accidentally sewed all the way through the cuff, place a piece of tape to insulate the connection (Figure 8.6).

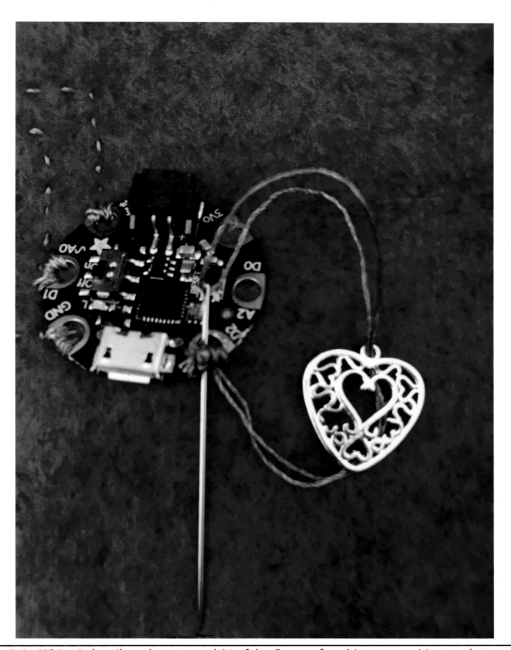

Figure 8.6   Whipstitch a silver charm to pad A1 of the Gemma for a bigger capacitive touch target.

Step 6: Print a Battery Case

Optionally, you can 3D print a battery sleeve with little sewing tabs that will protect the battery from accidental harm (Figure 8.7). The model is downloadable from the book website or available online.

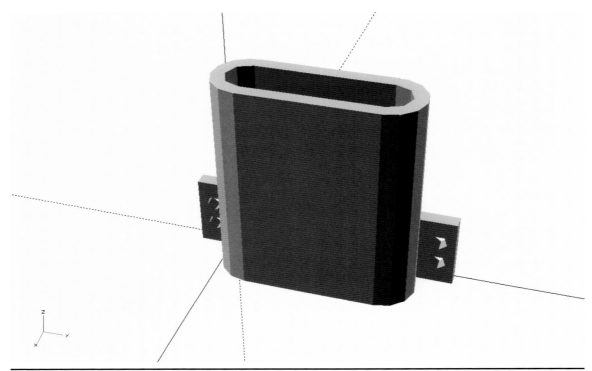

Figure 8.7   This small 3D-printable battery case can help project your LiPo battery from harm.

Step 7: Going Further

In this project, we used two of the three available input-output (IO) pins, leaving the third pin available to add a sensor or an actuator to your project. Another possibility is adding a small charger breakout board between the battery and the Gemma so that the battery can be charged from a USB cable directly on the cuff without having to remove the battery.

Step 8: Troubleshooting

If your cuff is not working properly, use the following steps to find the problem:

1. Is the status indicator LED on? If the indicator light is not lighted, check that the power slide switch is in the "on" position. If the power switch is on but the indicator LED is not lighted, check the battery. Unplug the battery, and plug it in to the battery charger to make sure that the battery is charged.

2. The status indicator LED in the program we have provided should light up green. This is the default behavior that indicates that a program is running. If the LED is flashing different colors, check the Gemma M0 troubleshooting guide online (https://learn.adafruit.com/adafruit-gemma-m0/troubleshooting).

3. If the Gemma does not show up as a USB drive on your computer but the status indicator light is green, check to be sure that the USB cable is not a USB power-only cable. Always remember to eject or dismount the Gemma M0 before unplugging it from a computer.

4. If one or more Flora NeoPixels are not lighting, check for loose connections. Make sure that your knots have not come undone and the whipstitching at the connection pads is tight.

5. If the circuit is behaving erratically, check that your stitches are not crossed. Make sure that your knots have not come undone and the ends are trimmed short. Seal the knots with Fray Check sealant or clear nail polish to prevent unraveling.

LED Matrix Purse

THE LED MATRIX PURSE BRINGS TOGETHER THE FABRICATION SKILLS YOU'VE LEARNED TO MAKE A fashion-forward and personalized accessory.

What You Will Learn

In this chapter, you will learn how to

- Work with an LED matrix
- Cut and assemble an acrylic box

Files You Will Need

- Laser cutter: casePlans.svg
- Laser cutter: mountingPanel.svg

Tools You Will Need

- Laser cutter
- Soldering iron and solder
- Hand drill or rotary tool (Dremel)
- Solvent cement applicator
- Hot-glue gun and glue sticks

Materials You Will Need

- 16 × 32 RGB LED matrix panel (www.adafruit.com/product/420)

- Pixel v2.5 Maker Kit (www.seeedstudio.com/Pixel-v2.5-maker-kit-p-2451.html)

- Lithium ion/polymer battery: 3.7 volts (V), 2,500 milliampere hours (mAh) (www.adafruit.com/product/328)

- USB to 2.1-millimeter male barrel jack cable (www.adafruit.com/product/2697)

- Adafruit PowerBoost 1000 Charger (www.adafruit.com/product/2465)

- Lithium ion/polymer battery: 3.7 volts (V), 2,500 milliampere hours (mAh) (www.adafruit.com/product/328)

- Coaxial 5.5/2.1-millimeter barrel connector: DC power plug (www.adafruit.com/product/3310)

- Hookup wire (preferably red and black, but any color will do)

- Micro SD card to SD card adapter

- 6 inches of double-sided or self-engaging hook and loop tape (Velcro) (www.velcro.com/business/products/self-engaging-hook-and-loop/)

- Acrylic sheet 3 millimeters ($\frac{1}{8}$ inch) thick, 16 × 12 inches or larger: gold or other glitters (A two-way mirror or other semitransparent acrylics are suitable for the purse case.)

- Acrylic sheet 3 millimeters ($\frac{1}{8}$ inch) thick, 8 × 4 inches or larger: black or any color for mounting panel

- Gloves and safety gear for working with acrylic cement

- Acrylic cement, solvent-based glue: SCIGRIP fast-set, clear, water-thin solvent cement

- Polyethylene squeeze bottle applicator with needle tip

- Polyethylene funnel for acrylic cement

- Two $\frac{3}{4}$-inch metal hinges

- 1-inch oxhorn or other metal clasp

- 12 screws and matching nuts to fit your hinges and clasps

- Four adhesive-backed bumpons for purse feet (0.14 inch high × 0.5 inch wide recommended)

- Blue painter's tape

The purse uses a mass-produced 512 RGB LED matrix, the same technology used to create video walls and large electronic signs. You'll create a custom laser-cut acrylic case and attach hinges and a clasp to house the electronics. The PIXEL maker board can be connected with bluetooth to change the art display or scroll text, and the board has an onboard micro SD card that saves LED art designs locally so that the art will loop indefinitely on the LEDs after your device has been Bluetooth disconnected (Figure 9.1).

The acrylic cement used to make the custom purse case contains toxic materials. Use gloves and carefully follow the safety warnings when using acrylic cement (Figure 9.2).

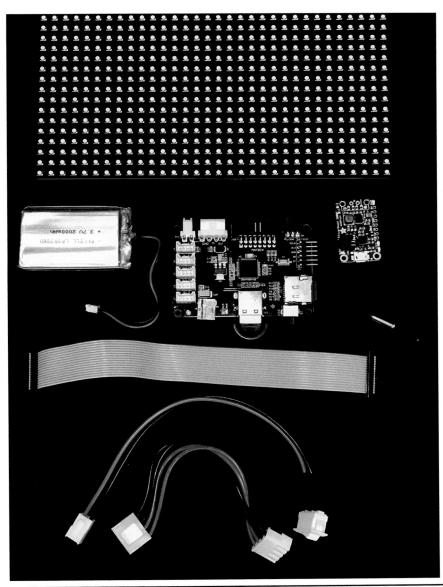

Figure 9.1 The electronic materials used in this project are the LED RGB matrix, a PIXEL Maker Kit, an Adafruit PowerBoost 1000, and a LiPo battery.

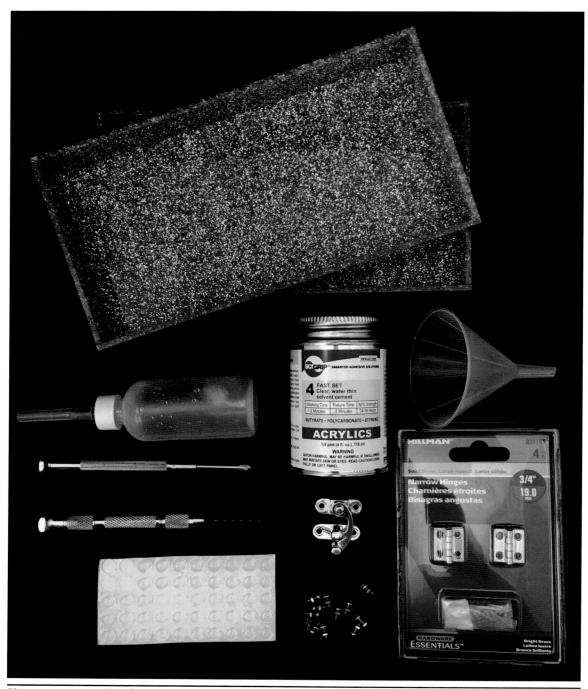

Figure 9.2   Supplies for the constructing the case include acrylic, solvent cement, a small funnel, and a small bottle applicator with needle tip. The hand drill is used to drill holes for the small screws and nuts used to attach the metal hinges and clasp.

Step 1: Laser Cut the Purse Case and Mounting Panel

1. Download the laser file for the acrylic purse case. The case is designed with box joints, also known as *finger joints*, for 3-millimeter-thick ($\frac{1}{8}$-inch) acrylic sheets. The file sizes are set for a laser working area of 18 × 24 inches. Depending on the sizes of your laser cutter and acrylic, you may need to rearrange the pieces of the case in a computer-aided design (CAD) program such as Inkscape or Adobe Illustrator.

2. Cut the eight pieces for the case.

3. Download the laser files for the mounting panel. Cut this panel using $\frac{1}{8}$-inch-thick acrylic in black or any other color. You will need to check the mounting hole placement of the 32 × 16 LED matrix panels because the panels will have different mounting holes depending on the manufacturer or run, and there is no mounting-hole standard. Compare the mounting holes layout of your 32 × 16 matrix against the design files, and modify if needed before laser cutting. Cut the mounting panel.

Step 2: Update the Firmware

1. The PIXEL Maker Kit is installed with an Android-only firmware. For this project, you'll need to switch the PIXEL firmware to one that is low power and supports both iOS and Android.

2. If you've plugged in the battery and switch, be sure to turn the PIXEL board off. Remove the micro SD card from your PIXEL board. Using an SD card adapter, plug the card into your computer.

3. Delete any existing files on the micro SD card.

4. The new firmware files are on the PIXEL maker webpage at http://ledpixelart.com/firmware/. Choose the iOS/Android 32 × 16 Firmware Image file, unzip it, and copy it to the micro SD card.

5. Insert the micro SD card back into your PIXEL board, connect the on/off switch, charge the battery, and power the PIXEL board on. Your firmware has now been updated.

Step 3: Prepare the PowerBoost 1000

1. Cut two strands of hookup wire about 20 centimeters in length, preferably one black and one red.

2. Solder the black wire to the longer outer barrel connector. Solder the red wire to the shorter inner tip connector (Figure 9.3).

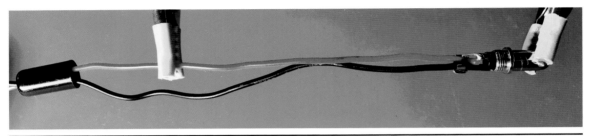

Figure 9.3   Solder wires to the barrel connector.

3. Solder the barrel jack wires to the GND and 5V pins of the PowerBoost 1000 near the PWR LED, black to ground and red to 5V (Figure 9.4).

Figure 9.4   Solder the barrel jack wires to the PowerBoost 1000.

4. Solder a power switch to the VS, EN, and GND pins (Figure 9.5).

Figure 9.5 Solder a power switch to the PowerBoost 1000.

Step 4: Attach the Electronics to the Mounting Panel

1. Add the PIXEL maker board to the mounting panel with $^{3}/_{8}$-inch 3-48 screws and nuts.

2. Using the screw holes in the corners of the Adafruit PowerBoost 1000, attach the PowerBoost to the plastic mounting panel with $^{3}/_{8}$-inch 3-48 screws and nuts (Figure 9.6).

3. Using a small amount of hot glue, attach the self-engaging hook and loop tape to the plastic mounting panel next to the PowerBoost 1000. This tape will hold the LiPo battery in place (Figure 9.7).

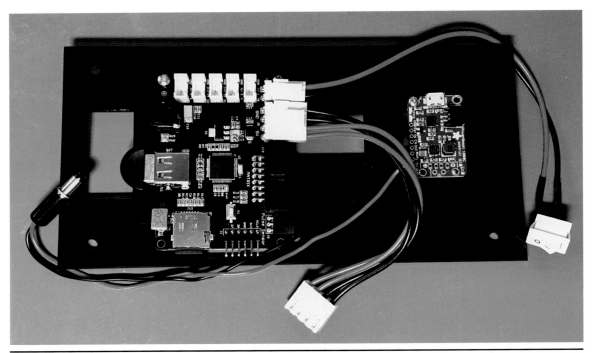

Figure 9.6   Use plastic or metal screws to attach the electronics to the mounting panel.

Figure 9.7   Use a small amount of hot glue to attach the self-engaging hook and loop tape to the plastic mounting panel next to the PowerBoost 1000. This tape will hold the LiPo battery in place.

4. Using snips cut off the two small plastic nubs on the frame of the LED matrix panel. This will allow the matrix frame to fit securely to the mounting panel (Figure 9.8).

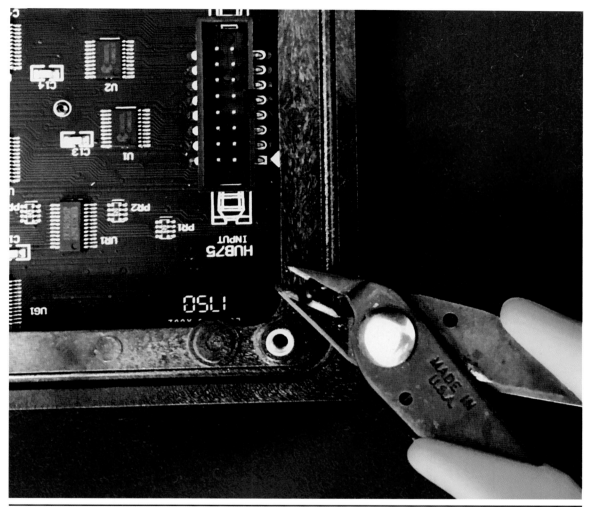

Figure 9.8   Snip the plastic nubs off the LED matrix frame.

5. Using the holes at the outer edges, attach the LED matrix to the mounting panel with the M3-50 screws.

6. Plug the PIXEL board in to the matrix using the power and control cables provided in the kit.

7. Plug the PowerBoost 1000 in to the board using the barrel jack.

8. Attach the battery to the PowerBoost 1000 (Figure 9.9).

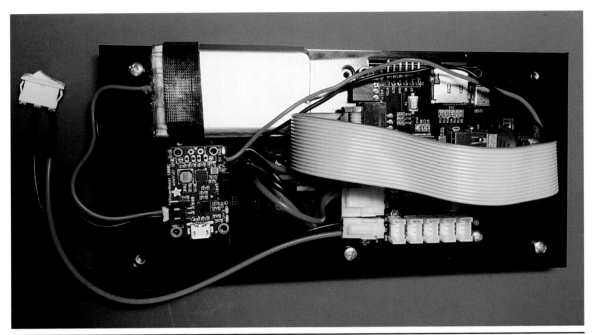

Figure 9.9   Attach the electronics to the mounting panel, and screw the panel onto the holes on the LED matrix.

Step 5: Assemble the Case

1.   Put on gloves and any other necessary safety gear, and use acrylic cement to assemble the halves of the case. Acrylic cement is a solvent that works by chemically welding the two pieces of acrylic together. This process happens quickly; we recommend that you watch some video tutorials and practice on spare acrylic before assembling your purse.

2.   Put two edges of acrylic together, and place the needle of the glue applicator where the pieces meet. Lightly squeeze a thin line of glue on the joint. Capillary action will spread the glue through the joint. Hold the pieces in place for several minutes using clamps or a jig (Figure 9.10).

Figure 9.10 Clamps hold the purse pieces in place for gluing.

3. Allow the glue to set for 10 to 15 minutes before moving.

4. Glue the four pieces on both sides of the purse. Allow the joints to cure for 24 to 48 hours for full strength.

5. Using blue painter's tape, tape the two halves of the case together. On the bottom of the purse, measure 1 inch in from the sides, and place one hinge evenly between the two halves of the case. Mark each of the screw holes onto the painter's tape (Figure 9.11).

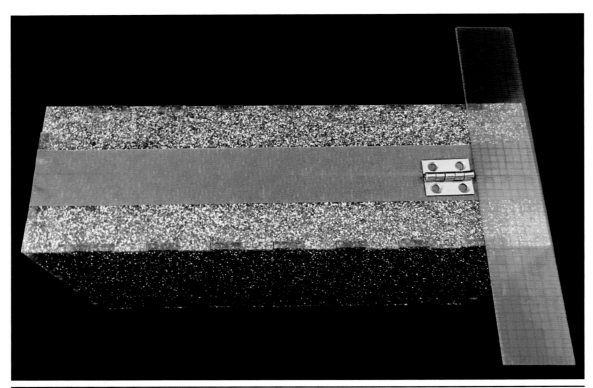

Figure 9.11 Mark the holes for the hinges and clasp.

6. Measure 1 inch from the other side of the bottom, and mark holes for the hinge.

7. On the top of the case, lay the clasp evenly between the two halves, and mark through holes onto the blue tape. Remove the hinges and clasp.

8. Using a hand drill or rotary tool (Dremel), drill holes on your marks for the hinges and clasp. (Figure 9.12)

Figure 9.12   Use a hand drill or rotary tool to make holes for the hinges and clasp.

9. Once you've drilled the holes, remove the blue painter's tape.

10. Place the hinges back on the outside of the purse, and use a screwdriver to insert the screws. Once the screws are in, open the purse, and add the matching nuts to the screws on the inside (Figure 9.13).

Figure 9.13   Use a screwdriver to attach the hinges on the outside of the purse.

11. Carefully slide the LED matrix and the electronics panel into the purse case until the LED matrix is flush with the acrylic case. The assembly should fit snugly in one half of the purse. You may need to remove and then replace the screws attaching the purse latch in order to fit the matrix and electronics panel.

12. Close the purse, place the clasp on the top, and using a screwdriver, insert the screws. Open the purse and add the matching nuts. Once you are done, be sure that your clasp latches closed.

13. On the bottom of the purse, where the hinges are, place four adhesive-backed bumpons on the corners as purse feet.

Step 6: Use It!

1. Download the CAT Clutch from either the iOS App Store (https://itunes.apple.com/us/app/c.a.t-clutch/id1038238338?mt=8) or the Google Play Store (https://play.google.com/store/apps/details?id=com.ledpixelart.wearable). This app is specific for the 32 × 16 matrix that is used in the purse.

2. With Bluetooth enabled on your phone, open the app to automatically pair with the purse.

3. With the CAT Clutch app you can play a variety of pixel art and animations. You can also scroll text using the Messages tab.

4. Create your own pixel art for the PIXEL board by making GIFs using Photoshop, Inkscape, or some other graphics programs (Figure 9.14). A tutorial for making pixel art for this project can be found at http://ledpixelart.com/for-artists/.

Figure 9.14 Make fun pixel art to display on your LED Matrix purse.

Index

Numbers

2D design
    downloadable files for this book, 29
    vector graphic files for laser, 24
3D design
    3D embellished t-shirt, 34–35
    CAD software drawings in, 24
    downloadable files for projects, 28
    printing. *See* 3D printing
    using Tinkercad, 26–27
3D embellished t-shirt
    3D print on fabric, 37–40
    assemble, 40–42
    create 3D design, 34–35
    cut out shirt pieces, 34
    export and slice model, 35–37
    materials needed, 32–33
    no-sew and low-sew options, 42–43
    what you will learn, files and tools, 31
3D printing
    3D design for, 26
    3D embellished T-shirt, 35–37
    battery and LED holder, 55
    Circuit Playground Express cover, 63
    as easy and affordable, 3–4
    lipo battery case, 138
    overview of, 25–26
    properties of fabric and, 32
    slicing 3D models with Cura for, 28
    fiber-optic holder, 113–114
    using Tinkercad, 26–27

A

acrylic, as laser cutting supply, 24
acrylic box, LED matrix purse
    acrylic cement and, 143
    assembling, 150–154
    laser-cutting, 145
acrylic cement, 143, 150–154
Adafruit
    Circuit Playground Express, 59, 61–63, 70
    as electronics source, 13
    Flora NeoPixel addressable LEDs, 129,
        132–136, 139
    Gemma M0, 129, 132–137, 139
    Micro Lipo Charger, 131
    microcontroller boards, 16, 22
    NeoPixel Library, 121–122
    PowerBoost 1000, 143, 146–147
adapter cable
    solar backpack preparation, 77
    starlight skirt electronics, 119–120
addressable LEDs
    NeoPixel. *See* Flora NeoPixel addressable
        LEDs
    overview of, 14
    sewable, 129
    starlight skirt electronics, 119–120,
        123–124
Adobe Illustrator
    design for laser with, 24
    fiber-optic fabric scarf, 48
    LED matrix purse, 145

anode, LEDs, 14
Arduino Desktop IDE
    downloading, 121
    programming Gemma M0 for cuff, 134
    starlight skirt, 122–123
    starlight skirt project, 102
Arduino Leonardo board, 123
Arduino microcontroller board
    addressable LEDs and, 14
    as electronic supply, 16–17
    overview of, 22
AutoCAD, design for laser, 24
awl, as sewing supply, 9

B

backstitch
    hand-sewing with conductive thread, 20–21
    with sewing machine, 9, 11
battery
    case for programmable sewn circuit cuff,
        138
    holder with lit LED for fiber-optic scarf, 55,
        56–57
    for solar backpack, 74, 80–81
    types used in this book, 15
Bluetooth, LED matrix purse and, 143, 155

C

CAD (computer-aided design)
    3D design for printing, 26
    design for laser, 24
    fiber-optic fabric scarf, 48
capacitive touch sensors, 136–137
capacitor, soldering solar charger to, 75–76
carbon dioxide (CO2) laser, 22
CAT clutch, LED matrix purse, 155
cathode, LEDs, 14
Circuit Playground Express microcontroller
    decorative cover for hip pack, 63, 70
    overview of, 59
    program for hip pack, 61–63
CircuitPython firmware, 22, 134–136
CO$_2$ (carbon dioxide) laser, 22
coin-cell battery
    3D fiber-optic fabric scarf, 46–47
    3D printing of, 55
    attach light source to, 56
    used in this book, 15

color
    addressable LEDs and, 14
    laser engrave files and differentiated, 131
    of polylactic acid filament for 3D printing, 26
computer-aided design. *See* CAD (computer-
        aided design)
computer, as electronic supply, 17
conductive thread
    as electronic supply, 15
    prepare leather for, 130
    sew electronics for programmable cuff, 133–134
    sewing with, 20–21
connectors
    JST-PH connectors, solar backpack, 78–80
    JST-SM plug connector, starlight skirt,
        119–120
    solar panel for solar backpack, 77–78
    solder USB for solar backpack, 78
    splice for solar backpack, 78–80
    starlight skirt electronics, 119–120
CorelDraw, design for laser, 24
cuff. *See* programmable sewn circuit cuff
Cura
    3D embellished t-shirt, 34–37, 39
    slicing with, 28
    starlight skirt, 113–114

D

data connections, programmable cuff, 133
driver, Circuit Playground Express, 62
dye, leather, 130

E

electrical tape, as electronic supply, 15
electronics
    LED matrix purse, 147–150
    overview of, 13
    programmable circuit cuff, 132–134
    starlight skirt, 119–120
    supplies and tools, 13–17
    techniques, 17–19
end-glow fiber optics, 16
Evil Genius series (McGraw-Hill), 18
exporting, 3D design for printing, 28

F

fabric scissors, 7
fabrics, laser cutting supplies, 24

festival hip pack. *See* fun festival hip pack

fiber-optic fabric scarf
    3D print battery and LED holder, 55
    attach light source, 55–57
    laser cut fabric for scarf, 48
    materials needed, 46
    no-laser option, 46–48
    split fiber-optic panel, 48–51
    sew fabric and attach to fiber-optic panel, 52–54
    what you will learn, files and tools needed, 45

fiber-optic panel/fabric. *See* fiber-optic fabric
        scarf

fiber optics
    as electronic supply, 16
    starlight skirt lit up with. *See* starlight skirt

file extensions
    2D vector design for laser, 24
    3D design, 28
    laser cutter, 28

Flora NeoPixel addressable LEDs
    laying out and sewing electronics, 132–134
    overview of, 129
    troubleshoot programmable circuit cuff, 139

fun festival hip pack
    3D print decorative cover, 63
    bring it all together, 70–71
    materials needed, 60–61
    overview of, 59
    program Circuit Playground Express, 61–63
    sew pack, 63–69
    what you will learn, files and tools needed,
        59–60

fused-filament fabrication, 3D printing, 25

G

Gamer Girl dresses, 1–2

Gemma M0 microcontroller
    lay out and sew electronics for cuff, 132–134
    overview of, 129
    program, 134–136
    troubleshoot, 139
    use capacitive touch sensor, 136–137

ground connections, electronics for
        programmable cuff, 133

H

hand-sewing stitches
    3D fiber-optic scarf, 49

with conductive thread, 20
    overview of, 10–11

helping hands stand, for electronics, 16

hip pack. *See* fun festival hip pack

I

IDE (Integrated Development Environment).
        See Arduino Desktop IDE

Inkscape, design for laser, 24

introduction, to wearable technology, 1–4

iron, as sewing supply, 9

J

JST-PH connectors, solar backpack, 78–80

JST-SM plug connector, starlight skirt, 119–120

K

Kapton tape, 75–76

L

laser cutter
    3D fiber-optic fabric scarf, 48
    as cutting supply, 24
    design for, 24
    downloadable files for, 29
    making cuff, 131
    overview of, 22–23
    working with leather, 130–131

leather
    circuit cuff. *See* programmable sewn circuit
        cuff
    needles, 9, 130
    as supply for laser cutting, 24
    working with, 130–131

LED diffusers, starlight skirt
    3D print, 113–114
    finished project, 125
    learning how to work with, 101
    program StitchKit MakeFashion board,
        121–124
    put fiber-optics together, 114–120

LED matrix purse
    assemble case, 150–154
    attach electronics to mounting panel,
        147–150
    laser cut purse case/mounting panel, 145
    materials for, 142–144
    prepare PowerBoost 1000, 146–147

LED matrix purse (*continued*)
    update firmware, 145
    use it! 155
    what you will learn, files and tools needed, 141
LED RGB matrix, 143
LEDs (light-emitting diodes). *See also*
    addressable LEDs
    3D fiber-optic fabric scarf, 46–47
    attach light source, 55–57
    battery holder, 55
    connect with smartphone app, 1–2
    as electronic supply, 14
    program Circuit Playground Express, 63
    solar backpack, 80–81
    video games on clothing with, 1–2
Library Manager, install NeoPixel from, 121–122
LiPo (lithium polymer) battery
    LED matrix purse, 143, 147–149
    programmable sewn circuit cuff, 131–132
    in wearable projects, 15
low-sew option
    3D embellished T-shirt, 42–43
    overview of, 5
    starlight skirt project, 102

M

MakeCode
    fun festival hip pack and, 59
    tutorials in, 59
    using with Circuit Playground Express, 61–63
    visual block programming with, 22
Make: magazine, 3D printer reviews, 26
MakeFashion Board, Stitchkit, 103, 120,
    121–125
manufactured fabric, as laser cutting supply, 24
materials
    3D embellished t-shirt, 32–33
    fiber-optic fabric scarf, 46
    fun festival hip pack, 60–61
    programmable sewn circuit cuff, 128–129
    solar backpack, 74
    starlight skirt, 102–103
measuring tape, as sewing supply, 9
micro:bit microcontroller board, 22
microcontroller boards
    Arduino, 14, 16–17, 22
    Circuit Playground Express, 59, 61–63, 70
    as electronic supply, 16–17

Gemma M0 microcontroller, 129–137
MakeFashion Board, Stitchkit, 103, 120,
    121–125
    overview of, 22
MicroPython programming language, 22
Mirror tool, Cura, 39–40
mounting panel, LED matrix purse, 145,
    147–150

N

naming conventions, Tinkercad, 27
natural fabric, as laser cutting supply, 24
needles
    3D embellished t-shirt, 32
    basic sewing, 8
    leather, 9, 130
    sewing machine, 9
NeoPixel Library, Adafruit, 121–124
NeoPixels. *See* Flora NeoPixel addressable
    LEDs
no-sew option
    3D embellished T-shirt, 42–43
    fun festival hip pack, 60
    overview of, 5
    solar backpack, 74, 82

P

patterns
    3D embellished t-shirt, 30, 32
    cutting out paper, 12
    downloadable for projects in this book, 28
    fiber-optic fabric scarf, 44
    fun festival hip pack, 59
    programmable sewn circuit cuff, 127, 132
    solar backpack, 73
Pause button, 3D printing, 35–36
photopolymerization, resin 3D printers, 25
photovoltaics. *See* solar backpack
pins, as sewing supply, 7
PIXEL maker board, LED matrix purse
    attach electronics to mounting panel,
        147–150
    connect to Bluetooth, 143
    creating own art for, 155
    update firmware, 143
PLA (polylactic acid) filament
    3D embellished t-shirt, 32
    festival hip pack, 60

starlight skirt, 102
use for projects in this book, 26
polyimide tape, 75–76
Ponoko laser cutting, 23
Post Processing Plugin, Cura, 37
Power Boost 500, solar backpack, 78, 80–81
PowerBoost 1000, LED matrix purse, 143, 146–147, 149
program files
  3D embellished t-shirt, 31
  downloading for projects in this book, 28
  fiber-optic fabric scarf, 45, 55
  fun festival hip pack, 60, 61, 63
  LED matrix purse, 141
  programmable sewn circuit cuff, 127, 134–136
  solar backpack, 73
  starlight skirt, 101, 113, 122
programmable sewn circuit cuff
  3D print battery case, 138
  charge battery, 131–132
  going further, 138
  lay out and sew electronics, 132–134
  make cuff, 131
  materials needed, 128–129
  program Gemma, 134–136
  use capacitive touch sensor, 136–137
  what you will learn, files and tools needed, 127–128
  working with leather, 130

R

resin printers, 3D printing, 25
resources, downloadable files for this book, 28
Rotate tool, Cura, 39
ruler, as sewing supply, 9
running stitch
  creating, 10
  sewing 3D fiber-optic fabric scarf, 48–54
  sewing conductive thread on cuff, 133

S

safety glasses, for soldering, 16
Scaled Vector Graphics (SVG) file, 34
scissors, fabric, 7
seam allowance, use standard 1/2," 11
seam ripper, as sewing supply, 8
Seeed Studio, 13, 16
sensor, as electronic supply, 16–17

sewing
  3D embellished T-shirt, 40–43
  basic supplies for, 8–10
  with conductive thread. *See* conductive thread
  fiber-optic fabric scarf, 48–54
  fun festival hip pack, 63–70
  hand-sewing stitches, 10–11
  programmable circuit cuff, 132–134
  solar backpack, 82–96
  starlight skirt, 103–112
sewing machine
  as needed supply, 9
  stitches, 11
sewn circuit cuff. *See* programmable sewn circuit cuff
Shapeways, 3D printing, 28
side-glow fiber optics, 16
silicone-covered wire, wearable projects, 15
slicing
  3D embellished T-shirt design, 35–37
  3Dprint fiber-optic holders/LED diffusers, 113–114
  with Cura for 3D printing, 28
smartphone app, connecting LEDs, 1–2
social media, wearable technology interacting with, 1–2
solar backpack
  attach solar panel to backpack, 97–98
  hooking everything up, 80–81
  make backpack, 82–96
  materials needed, 74
  prepare solar charger, 75–76
  prepare solar panel, 77–78
  solder USB connector on Power Boost 500, 78
  splice two JST-PH connectors, 78–80
  use it outside, 99
  what you will learn, files and tools needed, 73
solar charger
  avoid leaving in very hot places, 74
  hooking up for solar backpack, 80–81
  preparing for solar backpack, 75–76
solar panel
  attaching to backpack, 97–98
  direct sunlight best for, 74
  hooking up for solar backpack, 80–81
  preparing for solar backpack, 77–78

soldering
  capacitor to solar charger, 75–76
  helping hands stand for, 16
  power switch to PowerBoost 1000, 146–147
  safety glasses, 16
  splicing two JST-PH connectors, 79–80
  through-hole, 18–19
  types of solder in this book, 15
  USB connector on Power Boost 500, 78
  wires together, 18–19
soldering iron, 15
solid-core wire, 15
SparkFun, 13, 16
starlight skirt
  3Dprint fiber-optic holders/LED diffusers,
    113–114
  connect electronics, 119–120
  finished product, 125
  hand-sewn, 10
  machine-sewn, 11
  materials needed, 102–103
  program StitchKit's MakeFashion board,
    121–124
  put fiber optics together, 114–119
  sew skirt, 103–112
  what you will learn, files and tools needed, 98
StitchKit Fashion Technology Kit, 102
StitchKit MakeFashion board, 102
STL file, exporting from Tinkercad, 35
straight stitch, with sewing machine, 9, 11
stranded wire, 15
sunlight, solar backpack and, 98
supplies/techniques/fabrication machines
  3D design for printing, 26
  3D printing, 25–26
  design for, 24
  downloadable files for this book, 29
  electronic supplies and tools, 13–17
  electronic techniques, 17–19
  electronics, 13
  hand-sewing stitches, 10
  laser cutters, 22–23
  laser cutting supplies, 24
  machine sewing technique, 11
  machine sewn stitches, 11
  microcontroller boards, 22
  patterns, 12
  sewing supplies, 7–9
  sewing with conductive thread, 20–21
  slicing with Cura, 28–29
  using Tinkercad, 26–27
SVG (Scaled Vector Graphics) file, 34

T

Teach Yourself Electricity and Electronics
    (McGraw-Hill), 18
Thingiverse
  3D design models, 26, 28
  3D embellished T-shirt design, 34–35
thread, as sewing supply, 8
through-hole soldering, 15, 18–19
Tinkercad
  creating 3D design, 26–27
  creating 3D embellished T-shirt design,
    34–35
  exporting 3D design for slicing, 28, 35–37
  LED matrix purse, 145

U

USB cable, Circuit Playground Express, 62
USB connector, soldering on Power Boost 500,
    78

V

Velcro, festival hip pack, 64–65

W

wearable technology
  introduction, 1–3
  projects overview, 4–5
whipstitch, 10
wire cutter, as electronic supply, 16
wire stripper, as electronic supply, 16
wires
  attaching solar panel to backpack, 97–98
  soldering together, 18–19
  through-hole soldering, 18–19
  used in this book, 15
work-plane dimensions, 3D design with
    Tinkercad, 27

Y

YouMagine, 26, 28

Z

zigzag stitch, with sewing machine, 9, 11